T0176946

Open and Toroidal Electrophoresis

Open and Toroidal Electrophoresis

Ultra-High Separation Efficiencies in Capillaries, Microchips, and Slabs

Tarso B. Ledur Kist
Department of Biophysics, Federal University of Rio Grande do Sul, Brazil

This edition first published 2020
© 2020 John Wiley & Sons Ltd

All rights reserved. No part of this publication may be reproduced, stored in a retrieval system, or transmitted, in any form or by any means, electronic, mechanical, photocopying, recording or otherwise, except as permitted by law. Advice on how to obtain permission to reuse material from this title is available at http://www.wiley.com/go/permissions.

The right of Tarso B. Ledur Kist to be identified as the author of this work has been asserted in accordance with law.

Registered Offices
John Wiley & Sons, Inc., 111 River Street, Hoboken, NJ 07030, USA
John Wiley & Sons Ltd, The Atrium, Southern Gate, Chichester, West Sussex, PO19 8SQ, UK

Editorial Office
The Atrium, Southern Gate, Chichester, West Sussex, PO19 8SQ, UK

For details of our global editorial offices, customer services, and more information about Wiley products visit us at www.wiley.com.

Wiley also publishes its books in a variety of electronic formats and by print-on-demand. Some content that appears in standard print versions of this book may not be available in other formats.

Limit of Liability/Disclaimer of Warranty
In view of ongoing research, equipment modifications, changes in governmental regulations, and the constant flow of information relating to the use of experimental reagents, equipment, and devices, the reader is urged to review and evaluate the information provided in the package insert or instructions for each chemical, piece of equipment, reagent, or device for, among other things, any changes in the instructions or indication of usage and for added warnings and precautions. While the publisher and authors have used their best efforts in preparing this work, they make no representations or warranties with respect to the accuracy or completeness of the contents of this work and specifically disclaim all warranties, including without limitation any implied warranties of merchantability or fitness for a particular purpose. No warranty may be created or extended by sales representatives, written sales materials or promotional statements for this work. The fact that an organization, website, or product is referred to in this work as a citation and/or potential source of further information does not mean that the publisher and authors endorse the information or services the organization, website, or product may provide or recommendations it may make. This work is sold with the understanding that the publisher is not engaged in rendering professional services. The advice and strategies contained herein may not be suitable for your situation. You should consult with a specialist where appropriate. Further, readers should be aware that websites listed in this work may have changed or disappeared between when this work was written and when it is read. Neither the publisher nor authors shall be liable for any loss of profit or any other commercial damages, including but not limited to special, incidental, consequential, or other damages.

Library of Congress Cataloging-in-Publication Data Applied for

HB ISBN: 9781119539407

Cover Design: Wiley
Cover Image: Courtesy of Tarso B. Ledur Kist

Set in 9.5/12.5pt STIXTwoText by SPi Global, Chennai, India

Printed and bound by CPI Group (UK) Ltd, Croydon, CR0 4YY

10 9 8 7 6 5 4 3 2 1

To Glauber, Rubens, Zi, and Jaqueline.

Contents

Preface

The main goal of this book is to present electrokinetic based separation techniques in a didactic and concise manner, making the content accessible to the broadest possible audience (students, users, and researchers). In addition, this book provides the exact expressions of separation efficiency, resolution, peak capacity, and many other performance indicators. Appropriate illustrations are presented, and the underlying fundamental phenomena and models are discussed in a logical order instead of in a chronological order. Chemical and physical insights are also developed by emphasizing the microscopic scenarios of the molecular processes involved.

Important mathematical relationships are given to aid users with their practical applications and theoretical studies of the dozens of separation modes performed in the three most common platforms: capillaries, microchips, and slabs. The two most important layouts of these three platforms are also described and compared: the conventional *open* layout, with an inlet and an outlet for the samples, and the *toroidal* or closed loop layout, where samples run in a *quasi*-continuous circulating mode until the desired resolution is obtained.

Motivation to write this book came from twenty years of teaching both undergraduate and graduate students, as well as from stimulating collaborations with so many bright and talented colleagues, to all of whom I am deeply grateful.

The author is thankful to Dr C. Heller, Professor J. H. Z. dos Santos, Professor C. L. Petzhold, Professor A. Manz, Professor Bingcheng Lin, Professor B. Gaš, Professor J. W. Jorgenson, and Professor H-P. Grieneisen for their helpful discussions and suggestions. Additionally, the author is grateful to E. J. M. Bradley for proofreading the manuscript. Finally, it must be mentioned that these people should not be blamed for the errors the author insisted on keeping in this book!

January, 2020

Tarso B. Ledur Kist
Porto Alegre

Acronyms

AE	Affinity electrophoresis
BGE	Background electrolyte
CBC	Cyclic band compression
EC	Electrochromatography
EDTA	Ethylenediaminetetraacetic acid
ELFSE	End-labeled free-solution electrophoresis
EOF	Electroosmotic flow
FFE	Field flow electrophoresis
FFF	Field flow fractionation
FSE	Free-solution electrophoresis
EST	Electrokinetic separation technique
ID	Inner diameter (of a capillary)
IEF	Isoelectric focusing
IEP	Isoelectric point
ITP	Isotachophoresis
MEEKC	Microemulsion electrokinetic chromatography
MEKC	Micellar electrokinetic chromatography
OD	Outer diameter (of a capillary)
SE	Sieving electrophoresis
Tw/CBC	Toroidal layout with cyclic band compression
Tris	Tris(hydroxymethyl)aminomethane or 2-amino-2-(hydroxymethyl)propane-1,3-diol

Symbols and Conventions

a	Capillary internal radius
a_i	Chemical activity of species i (dimensionless quantity)
A	Cross sectional area of the separation medium
α	Band compression factor
α	Degree of dissociation of a solute
b	Injected sample plug length
B	Band capacity
B_t	Band capacity per unit time (band rate)
b_i	Concentration of component i in molality (mole/Kg of solvent)
b°	Standard state of solute concentration given in molality
β	Buffer capacity, $\beta = dn/d(\mathrm{pH})$
c_i	Amount concentration of component i in molarity (mol/L of solution)
c°	Standard state of amount concentration given in molarity
c_{tot}	Sum of amount concentration of species of an acid or base
D	Molecular diffusion coefficient
D_{ads}	Dispersion coefficient due to wall adsorption
D_{coil}	Dispersion coefficient due to capillary or microchannel coiling
D_{ieof}	Dispersion coefficient due to inhomogeneous EOF
D_{emd}	Electromigration dispersion coefficient due to non-uniform BGEs
D_{pvf}	Dispersion coefficient due to parabolic or Poiseuille velocity flow
D_{temp}	Dispersion coefficient due to radial temperature gradients
ΔG	Gibbs free-energy change
ΔH	Height difference between reservoirs
ΔH	Enthalpy change
ΔP	Pressure difference between reservoirs
ΔS	Entropy change
e	Elementary charge, $e \simeq 1.6 \times 10^{-19}\,\mathrm{C}$ (a real and positive number)
E	Applied electric field strength
ϵ_{r}	Relative (static) permittivity

$\epsilon_\circ$	Electric constant (vacuum permittivity)
η	Dynamic viscosity of a liquid (solvent or solution)
f	Viscous friction coefficient
F	Faraday constant
F_e	Electric force, $F_e = q\,E$
F_f	Friction force, $F_f = f\,v$
$\gamma_{x,i}$	Activity coefficients for concentration unit x of species i (dimensionless quantity)
h	Half the height of a microchannel or slab
H	Height equivalent of a theoretical plate
i	Direct electric current, $i = dQ/dt$
I	Ionic strength, amount concentration basis, $I = \frac{1}{2}\sum_i c_i z_i^2$
j	Direct electric current density, $j = i/A$
κ	Reciprocal of double-layer thickness
κ	Direct current conductivity
k	Heat conductivity $(\mathrm{Wm^{-1}K^{-1}})$
K_a	Ionization equilibrium constant of an acid
K_a^*	Acid ionization constant when at an electronically excited state
k, k_B	Boltzmann's constant
K'_{AS}	complexation or ligation apparent equilibrium constant
l	Length from injection point to detection point
L	Separation medium total length in the open layout ($E = V/L$)
L	Half the distance of a full turn in the toroidal layout (E is also given as $E = V/L$ in this case)
m	Mass of an entity (ion, molecule, particle)
μ	Apparent electrical mobility (with respect to the platform wall, $\mu = \mu_{el} + \mu_{eo}$). This is by far the most used in this book, therefore the privilege of carrying no index is given to it.
μ_e	Electrophoretic mobility (with respect to the buffer solution. The result of the electric and friction forces only, $\mu_e = q/f$)
μ_{ef}	Effective electrophoretic mobility of a molecule with respect to the buffer solution, $\mu_{ef} = \alpha\,\mu_e$, where α denotes the degree of ionization of the solute
μ_{el}	Electrical mobility (with respect to the buffer solution and regardless of the underlying phenomena, i.e. presence or absence of ligands, sieving agents, micelles, microemulsions, or any other pseudophase)
μ_{eo}	Electroosmotic mobility, $\mu_{eo} = v_{eo}/E$
$\mu_\circ$	Limiting eletrophoretic mobility (physical constant usually found in standard tables, determined when the analyte is at full charge and infinite dilution)
n	Number of equivalents of an acid or base

n	Number of half turns performed in a closed loop toroidal path
N	Number of theoretical plates (plate number or separation efficiency)
N_t	Number of theoretical plates produced per unit time (plate rate)
$\mathcal{N}$	Number of theoretical plates produced per unit time squared (plate double-rate or time efficiency)
N_A	Avogadro's constant ($6.02214076 \times 10^{23}$)
p	Heat generated by the Joule effect per unit volume, $p = \rho\,j^2$
P	Peak capacity
P_t	Peak capacity per unit time (peak rate)
P	Pressure
pH(I)	The value of the isoelectric point of a molecule
$\Psi_\circ$	electric potential at a dielectric–liquid interface
q	Net charge of an ion (time or ensemble average)
r	Cylindrical spatial coordinate (r, θ, x)
R	Gas constant ($R = N_A k_B$)
$R_{i,j}$	Resolution between the bands or peaks i and j
R_t	Resolution produced per unit time (resolution rate)
ρ	Direct current resistivity, $\rho = R\,L/A$
ρ	Density of a liquid or solution
ρ_i	Concentration of component i in mass per volume basis
ρ°	Standard state of solute concentration in mass per volume basis, e.g., g L^{-1} of solution
s°	Unit of detector's signal (1 mAbs, 1 μV, 1 μA, or other)
σ	Surface charge density, $\sigma = Q/A$
σ	Standard deviation of a band or peak
σ^2	Cumulative peak variance from all sources
σ^2_{ads}	Variance due to wall absorption
σ^2_{coil}	Variance due to capillary or microchannel coiling
σ^2_{det}	Variance due to detection width
σ^2_{D}	Variance due to molecular diffusion
σ^2_{ieof}	Variance due to inhomogeneous EOF
σ^2_{pvf}	Variance due to parabolic or Poiseuille velocity flow
σ^2_{temp}	Variance due to radial temperature gradients
t	Time coordinate
t°	Unit of time (1 min, 1 s, or other)
t_{eo}	Electroosmotic hold-up time
t_m	Migration time of the center of mass of the analyte band or peak
t_{mc}	Migration time for a micellar aggregate
T	Temperature
T_c	Temperature applied to the cyclic compression segment of the toroid
T_r	Running temperature

τ	Migration time of half a turn along the centerline of a toroid
θ	Cylindrical spatial coordinate (r, θ, x)
v	Apparent velocity (with respect to the platform wall, $v = v_{el} + v_{eo}$)
v_e	Electrophoretic velocity (with respect to the buffer solution. The result of electric and friction forces only, $v_e = qE/f$)
v_{ef}	Effective electrophoretic velocity of a molecule with respect to the buffer solution, $v_{ef} = \alpha\, v_e$, where α denotes the degree of ionization of the solute
v_{el}	Velocity due to electrical mobility (with respect to the buffer solution and regardless of the underlying phenomena, i.e., presence or absence of ligands, sieving agents, micelles, microemulsions, or any other pseudophase)
v_{eo}	Electroosmotic flow velocity, $v_{eo} = \mu_{eo}E$
v_o	Limiting eletrophoretic velocity (with respect to the buffer solution and determined when the analyte is at full charge and infinite dilution)
V	Applied electrostatic potential difference, $V = V_+ - V_-$
V_i	Electric potential at position i
V_+	Electric potential at the anode (anion destination)
V_-	Electric potential at the cathode (cation destination)
ω	A variable (set of numbers) simulating white noise
w	Baseline peak width
w	Half width of a rectangular microchannel or slab
x	Cartesian spatial coordinate (x, y, z), direction of run
x	Cylindrical spatial coordinate (r, θ, x), direction of run
x^o	Unit of distance (1 cm, 1 mm or other)
y	Cartesian spatial coordinate (x, y, z), orthogonal to the wall
z	Number and sign of elementary charges ($z = 0, \pm 1, \pm 2, ...$)
z	Cartesian (x, y, z) spatial coordinate, orthogonal to the wall
ζ	Electrokinetic potential or "zeta" potential (electrostatic potential at the shear plane or slipping plane)

Introduction

Toroidal electrokinetic separation techniques are based on separation tracks with toroidal layouts. These techniques can produce analytical and preparative separations with unprecedented high resolutions and peak capacities. Runs are performed in a closed loop with a quasi-continuous circulating mode of migration until the desired resolutions are achieved. They are different to the commonly used open layouts, where runs are always limited in either space (with an inlet and an outlet) or time (with a start and a finish line) and either electroosmosis or a pressure driven counter-flow must be applied to increase the resolving power. Toroidal layouts allow much more freedom in the use of the operating conditions, as will be described and compared in the following chapters.

Electrokinetic phenomena is a class of phenomena that includes electrophoresis, electroosmosis, streaming current and potential, surface conductivity, dielectric dispersion, and electroacoustics. The phenomena occur in liquid solutions (some of them can also occur in gels) containing electrolytes, and are intimately related to the *theory of electromagnetism* and *classical fluid dynamics*. However, they are distinct from electrochemistry related phenomena as they focus on the transport of charged and uncharged entities instead of on the chemical reactions, which involve the movement of electrical charges between electrodes and electrolytes.

The application of static or alternating electric fields to liquid solutions promotes the analytical and preparative separation of charged and uncharged entities, ranging from mono-atomic to macroscopic particles. This has led to the development of dozens of techniques that are indistinctly called electrophoretic, electromigration, electromigrative, electrodriven, or electrokinetic separation techniques. Luckily, all of them have the same acronym: ESTs! However, the term electrophoretic seems to be too specific, as it only refers to one of the electrokinetic phenomena. The term electromigration became commonly used in microelectronics to denote the phenomenon of atom displacement in solid conductors caused by the flow of electrons (the term "displacement of atoms by the electron wind" is commonly used in this field). Therefore, this word does not appropriately describe the separation techniques discussed here. The terms electrodriven and electromigrative are, perhaps, too broad to be used to refer to these separation techniques. In conclusion, the term *electrokinetic separation techniques* seems to be the most precise name for these techniques, and will hereafter be taken as their official name, as it undoubtedly specifies the set of phenomena used in the separations.

The categorization of the dozens of electrokinetic separation techniques (ESTs) is highly necessary as they have important theoretical and practical differences. Consequently, the categorization of these techniques into three levels is proposed as it produces a simple and practical nomenclature. This categorization begins with the first level: the layout of the separation path, which can be either open (O) or toroidal (T) in shape (as previously described).

The second level of categorization is the *platform* where the ESTs are performed. Capillary (C), microchip (M) and slab (S) platforms are the most common. Flexible fused-silica microtubes, popularly known as capillaries, are already widely used to achieve high separation efficiencies in the open layout and are also starting to be used in the toroidal layout. Cylindrical capillaries are the most frequently used, but square and rectangular capillaries are also available. In the case of microchips, microchannels are etched onto slides of polymeric materials, glasses, fused silica (amorphous), and quartz (crystalline). They show great promise for the development of novel, multifunctional microstructures. The use of 3D printing allows an even larger number of techniques to be performed on this platform, both for basic research and an uncountable number of applications. The slab platform is normally made of a slab of gel, cellulose acetate, nylon membrane, or another porous substance. These macroscopic supports always have an anti-convective effect that prevents the sample from spreading due to convection. In addition, they usually play additional, well established, roles in the separation mechanisms.

The third and lowest level of categorization is the *separation mode*. This refers to the underlying molecular mechanism used to promote the separation of the analytes of interest, both among themselves and from any undesired sample interferant. The modes include affinity electrophoresis (AE), electrochromatography (EC), end-labeled free-solution electrophoresis (ELFSE), free-solution electrophoresis (FSE), isoelectric focusing (IEF), isotachophoresis (ITP), microemulsion electrokinetic chromatography (MEEKC), micellar electrokinetic chromatography (MEKC), and sieving electrophoresis (SE), to mention only a few. It is interesting to note that this nomenclature is currently normally used for the separation modes, as a result of the recommendations made by IUPAC.

When combining these three levels of categorization to accurately name, produce an acronym, and describe an EST, the following intuitive rules are proposed: (1) the first word represents the first level of categorization; either open (O) or toroidal (T). (2) The second word specifies the platform: capillary (C), microchip (M), or slab (S). (3) The last word specifies the separation mode. A summary of the ESTs covered in this book is given in alphabetical order in the following table using this proposed nomenclature.

Layout	Platform	Separation mode
(1) Open	(1) Capillary	(1) Affinity electrophoresis
(2) Toroidal	(2) Microchip	(2) Electrochromatography
	(3) Slab	(3) End-labeled free-solution electrophoresis
		(4) Free-solution electrophoresis
		(5) Isoelectric focusing
		(6) Isotachophoresis
		(7) Microemulsion electrokinetic chromatography
		(8) Micellar electrokinetic chromatography
		(9) Sieving electrophoresis

The full names and acronyms of these 54 combinations are presented in Appendix A. However, the feasibility of some of these techniques is not obvious. For instance, running the isotachophoresis mode in the toroidal layout may be difficult because the leading and terminating electrolytes must be exchanged before each high voltage rotation (see Chapter 6).

Free flow electrophoresis (FFE) is a special layout mainly used for preparative separations in special macro- and microchip platforms (shallow two-dimensional chambers). Many separation modes can be performed in this setup. The sample is continuously fed as a narrow streak into the stream of the separation medium and an electric field is orthogonally applied to this flow direction, driving the components of interest to specific collection points of the outlet. This is also discussed in Appendix A.

It must be remembered that, for some applications, gel slabs can be used for at least two purposes; merely as an anticonvective agent (to prevent band spreading) and/or as a sieving agent. In the latter case a high rate of analyte collisions against the polymer net is observed and plays a fundamental role in the separations. All of these techniques and their respective underlying phenomena are discussed in relation to the open and toroidal layouts throughout this book.

Electrophoresis and electroosmosis are the most commonly used phenomena that drive separations in the ESTs. Moreover, a large quantity of additional basic phenomena can be directly observed and studied in the many ESTs. Some examples of these additional phenomena include fluid flow, analyte adsorption, phase partitioning, and the degree of acid–base ionization, as well as the interactions of ion–ion, ion–molecule, target–ligand, antibody–antigen, and many others. Moreover, the quantity of phenomena that can be mathematically modeled using well established theories is remarkable. Most of these phenomena can be modeled from first principles, without the need to add empirical or ad hoc parameters. Even so, as in all areas of science, there are a few experimental observations for which good models do not currently exist, and some of them even lack an adequate theory. It is an interesting field for developing models, testing theories, making predictions, and testing hypotheses involving quantifiable predictions and measurements. Finally, but no less importantly, the resulting separation techniques have an impressive number of applications within diverse fields such as clinical analysis, pharmaceutical analysis, genetic analysis, food analysis, environmental analysis, and proteome analysis, to mention only a few.

The present book is written in such a way as to – hopefully – make it an interesting read for experts in the field as well as for users of these technologies, non-specialists, and students. The whole book is richly illustrated and presents a large number of very useful equations showing the relationships between important operational parameters and other fundamental variables. These are important tools for both readers who are interested in the theories of the field and those who are interested in the practical applications of the ESTs. Mathematical deductions are shown only in the appendices because the intent is not to bother readers who are not very familiar with mathematical methods. Students interested in becoming familiar with the modeling of the electrokinetic phenomena are encouraged to read these appendixes.

In Chapter 1 the fundamental concepts and definitions related to liquids that may contain a background electrolyte are reviewed, paying specific attention to their applications to the ESTs. Phenomena and definitions studied in this chapter include the relative permittivity of water, dissolution, solvation, dissociation, ionization, Gibbs free energy, ionization constants, both c-pH and pa-pH diagrams, the Henderson–Hasselbalch equation, and the buffer capacity of aqueous solutions.

Chapter 2 is dedicated to the fundamentals of ESTs. This includes the electrophoresis of single molecules, ionic limiting mobility, bands, peaks, zones, isoelectric points, both turbulent and laminar flow, electroosmosis, suppression of electroosmosis, the Joule effect, heat dissipation, temperature profiles, molecular diffusion, band broadening, sample stacking, band compression, and the separation modes.

The focus of Chapter 3 is the open (common) electrophoresis layout. The relationships between independent variables (or users' operational parameters) and the performance indicators are given for this type of layout (open). Examples of such performance indicators are: number of theoretical plates, number of theoretical plates per unit time squared, plate height, resolution, resolution per unit time, peak capacity, band capacity, and both peak and band capacity per unit time.

In Chapter 4 the toroidal layouts of the three platforms (capillary, microchip, and slab) are presented in detail. The microholes (capillary), microconnections (microchip), and connections (slab) that function as the hydrodynamic and electrical communication between the toroid's internal lumen and the external environment (reservoirs and electrodes) are examined in detail. The concept of the passive and active modes of operation are also presented. Active modes are used to prevent the bands from leaking out of the toroids and into the reservoirs.

Chapter 5 gives a summary of the performance indicators presented in Chapters 3 and 4. Tables comparing the performance indicators as functions of the operational parameters are shown in this chapter. The performance indicators of the open and toroidal layout are contrasted and the pros and cons of each layout are examined.

The high voltage setups used in the open and toroidal layouts are discussed in Chapter 6. One important aspect that differentiates the toroidal layout from the open layout is the way the high voltages are connected and operated. Conventional positive and/or negative high voltage modules can be used; however their output must be quickly redistributed (rotated) in a cyclic manner to keep the set of bands running until the desired resolutions are achieved. This is performed using high voltage distributors and the pros and cons of each are shown using didactic illustrations.

Heat removal and temperature control in both open and toroidal layouts are presented in Chapter 7. In addition to the side effects of the temperature gradients on band dispersion, many unexploited potentials and advantages of a rational cooling design are presented and discussed for all platforms (capillary, microchip, and slab). This is examined with the aid of dozens of figures and a solid theoretical basis.

Most of the detectors that are compatible with the open layout are also compatible with the toroidal layout. They are presented in Chapter 8, which also includes the following detection systems: absorption, contactless conductivity, fluorescence, thermal lens, and mass spectrometry. Many advantages of fluorogenic and labeling reactions in both the open and toroidal layouts are presented. In addition, the use of these reactions to increase the detection limits, performance indicators, and separation selectivities is also discussed.

In Chapter 9 a few examples of the applications of the toroidal layout are presented. These include the analyses of amino acids, stereoisomers, and isotopomers, among others. The potential of toroidal electrophoresis to obtain separation efficiencies of over one hundred million theoretical plates is shown and discussed. This is the only chapter that omits the open layout, as it would be almost impossible to present all of its applications (across all platforms and separation modes) in a single book.

Important mathematical deductions pertaining to the ESTs are made in the appendixes. These deductions have been left for the appendixes to make the main text more fluidic and light.

1

Solvents and Buffer Solutions

1.1 Water as a Solvent

Water is by far the most widely used solvent in the preparation of gels, linear polymer solutions, and buffer solutions for electrokinetic separation techniques (ESTs). This is because of its unrivaled capacity to dissociate and ionize a great variety of substances (salts, acids, and bases). Ultimately, due to the large number of reactions that occur within water, it is the substance that makes life possible. Therefore, there are many reasons to take a closer look at the interesting properties of this unique liquid.

1.1.1 Temperature and Brownian Motion

The temperature of a solvent, for instance water, acetonitrile or methanol, is related to the average energy per degree of freedom that these molecules have. Average energy is given by $\langle E \rangle = k_B T$, where k_B represents the Boltzmann constant and T represents temperature in Kelvin.[1] The degrees of freedom consist of the translation velocity, which occurs in the three space coordinates, and both the vibrational and rotational degrees of freedom of the molecules. Raising the temperature of liquid water at 1 atm from 278.15 K (5 °C or 41 °F) to 348.15 K (75 °C or 167 °F) increases the average velocity (v), indeed $\sqrt{\langle v^2 \rangle}$, by about 25%. The consequential increase in thermal energy is important for the collision rate and momentum transfer among the molecules within the liquid. Such collisions with fast-moving molecules cause random movements of the microscopic particles that are suspended in the liquid (or in a gas), which can be observed under the microscope (*Brownian motion*). This is the basis for the denaturation and renaturation of nucleic acids, as well as the denaturation of proteins (disruption of both the secondary and tertiary structures). This same process also plays a role in dissolution, dissociation, ionization, and in maintaining the equilibrium of reactions. The Gibbs free energy change quantitatively gives the role of enthalpy, temperature and entropy for chemical reactions in the general case (see Section 1.1.8).

Open and Toroidal Electrophoresis: Ultra-High Separation Efficiencies in Capillaries, Microchips, and Slabs, First Edition. Tarso B. Ledur Kist.
© 2020 John Wiley & Sons Ltd. Published 2020 by John Wiley & Sons Ltd.

1.1.2 Electric Permittivity of Water

Water is a remarkable solvent for the dissolution, dissociation, and ionization of many substances. There are a few ways to predict these unique properties of water. The predictive models belong either to the classical theory of electromagnetism, statistical mechanics, thermodynamics, or the molecular dynamics simulations that use dedicated software to predict the physical movements of atoms and molecules (which are based on the fundamental Newton equations of mechanics). Above all, it must be remembered that oxygen has the second highest electronegativity of the whole periodic table [2] (see Figure 1.1). Only fluorine has a higher electronegativity; it forms only one *covalent bond* with a hydrogen atom (hydrogen fluoride) and is not liquid at room temperature (above 19.5 °C). Nitrogen, the other neighbor of oxygen in the periodic table, forms three covalent bonds with hydrogen and undergoes sp^3 hybridization, producing three sp^3 bonding orbitals and one sp^3 non-bonding orbital (forming ammonia). It is also a gas at room temperature. In water molecules, on the other hand, oxygen establishes covalent bonds with two hydrogen atoms. The resulting sp^3 structure contains two bonding orbitals (with a partial positive charge at each hydrogen position) and two non-bonding orbitals (with a partial negative charge on each). This produces a large electric dipole moment, especially considering that this is a small molecule, which explains the high relative electric permittivity of water (a small molecule with a large electric dipole moment), as shown in Table 1.1. The large amount of hydrogen bonds per unit volume within bulk water plays an important role in many of its unique properties. For example, the relatively high values of surface tension, boiling point, thermal conductivity, and latent heat of evaporation, to name a few. The strength and directional feature of hydrogen bonds explains a few additional odd properties, e.g., the solid (ice) has a lower density than the liquid (water).

The relative electric (static) permittivity of water is represented by the constant ϵ_r and has an important role in the *Coulomb equation*, equation 1.1. This constant is high for water and Table 1.1 shows a list for comparison. Moreover, the average number of *hydrogen bonds* between water molecules in pure water is much larger than the average number of hydrogen bonds within hydrofluoridic acid or ammonia, due to the symmetry of the two sp^3 bonding orbitals and the two sp^3 non-bonding orbitals of water molecules. This maximizes the average number of intermolecular hydrogen bonds – up to two as negative charge donor and two as negative charge acceptor – totaling up to four hydrogen bonds per molecule. Therefore, pure water is liquid from 273.15 K up to 373.15 K at 1 atm. The electric force (F_e) among ions within a liquid is given by:

$$F_e = \frac{1}{4\pi\epsilon_r\epsilon_o} \frac{q_1\,q_2}{d^2} \quad \text{with} \quad \epsilon_r = \frac{\epsilon}{\epsilon_o}, \tag{1.1}$$

where ϵ_o represents the electric permittivity of the vacuum, ϵ represents the permittivity of the liquid (solvent), ϵ_r is defined as the relative permittivity, q_1 and q_2

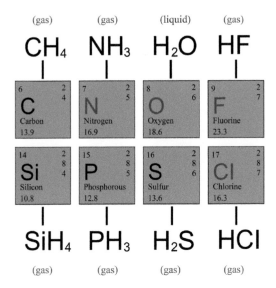

Figure 1.1 Some neighbors of oxygen in the periodic table. The atomic number is at the upper left, the electronic configuration at the upper right, and the electronegativity [2] at the bottom left. Oxygen has the highest electronegativity of all, except for fluorine. The hydrides shown are: methane, ammonia, water, hydrogen fluoride, silicon hydride, phosphine, hydrogen sulfide, and hydrogen chloride. All are gases in the 0 to 100 °C range and one atmosphere of pressure, except for water and hydrogen fluoride, which has a boiling point of 19.5 °C.

are the charges, and d the distance from each other. Table 1.1 shows the relative (static) permittivity of some solvents; note that water is among the highest.

1.1.3 Dissolution

Neutral and non-ionizable solutes that are able to form hydrogen bonds tend to be soluble in water. Of course, solubility depends on the size of the molecule, the number of hydrogen bonds each molecule forms with water, the interaction energies of the solute molecules with themselves, and the entropy changes involved. If the number of hydrogen bonds is small and the solute molecules are large compared to the water molecules, then the solute tends to be expelled out from the bulk (interior) of the water by the water molecules. An exact prediction of solubility is given by the Gibbs free energy change (see Section 1.1.8).

1.1.4 Solvation

The introduction of ions into certain solvents makes some solvent molecules attach around these ions, producing one or more layers of solvent molecules. The number of layers mainly depends on the charge of the ions, while the number of

Table 1.1 Relative (static) permittivities of liquids at 20 °C. Extracted from Wohlfart and Lechner (2008) [3].

Liquid	ϵ_r
N-methylformamide	189.0
Formamide	111.0
N-ethylformamide	104.7[a]
Water	80.100
Propylene carbonate	66.14
Methanoic acid (Formic acid)	51.1
Dimethyl sulfoxide	47.24
Glycerol	46.53
Ethylene glycol	41.4
N,N-dimethylacetamide	39.0
N,N-dimethylformamide	38.25
Acetonitrile	36.64
1,3-propanediol	35.1
Methanol	33.0
Ethanolamine	31.94
N,N-diethylformamide	29.6
2-pyrrolidone	28.18
Diethanolamine	25.75
Ethanol	25.3
Propan-2-one (Acetone)	21.01
1-propanol	20.8
2-propanol	20.18
Allyl alcohol	19.7
1-butanol	17.84
2-butanol	17.26
Cyclohexanone	16.1
Cyclopentanone	13.58
Trifluoroacetic acid	8.42
Tetrahydrofuran	7.6
Ethanoic acid (Acetic acid)	6.20
Propanoic acid	3.44
Dimethyl carbonate	3.087
1,4-dioxane	2.2189
Cyclohexane	2.0243
Hexane	1.8865

a) Calculated considering the change of 0.45% per °C observed with the formamide derivatives at 20 °C.

Figure 1.2 Schematic representation of chloride anion solvation in water.

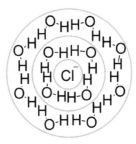

Figure 1.3 Solvation of the sodium cation in water.

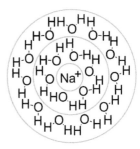

water molecules per layer largely depends on the size of the ions. These layers are a result of the *charge–permanent dipole interaction* between the ions and the permanent dipoles of the solvent molecules. Since water is a small molecule with a large electric dipole moment, this interaction is very strong in water. Figure 1.2 illustrates the solvation of an anion in water and Figure 1.3 illustrates the solvation of a cation in water.

1.1.5 Dissociation

Ionic liquids and ionic solids are predominantly made by cations and anions that are held together by electrostatic attraction. This is the state of salts and even some bases and acids. They tend to stay in this state when mixed with most solvents at room temperature. The thermal motion of the solvent molecules, even at room temperature, continuously pushes ions from the edges and corners of these ionic solutes due to inelastic collisions. In most solvents the electric restoring force, given by equation 1.1, pulls the the ions back. Within water, however, two things happen:

1) The detached ions are immediately subjected to solvation
2) The restoring force is much weaker because the relative permittivity (ϵ_r) of water is very high (equation 1.1).

These two properties of water make it a very good solvent for the dissociation of ionic liquids or ionic solids at room temperature.

As well as looking at the forces involved, another way to understand the unique ability of water to dissociate ionic liquids and solids is to examine the energies involved. Again, suppose that an ion is momentarily separated by thermal energy from a vertex or edge of a small neutral crystal stone of a solute. The detached ion is then immediately solvated and is initially positioned at distance a, in close contact with the crystal. The energy necessary to move this ion from this position ($x = a$) to infinity is found by replacing d by x in equation 1.1 and integrating it over dx, from $x = a$ to infinity. The result is given by:

$$U_e = \frac{1}{4\pi\epsilon_r\epsilon_o} \frac{q^2}{a} . \tag{1.2}$$

This is the energy necessary to move the ion with charge q, and the resulting oppositely charged small crystal with charge $-q$ from $x = a$ to infinity. From Table 1.1 is possible to see that in water this required energy is 10 times smaller than what would be necessary, for instance, to perform the same operation in tetrahydrofuran, which has a relative permittivity of 7.6. All of this was calculated without counting the solvation phenomena, which favors water over most other solvents. In conclusion, both the electric forces among ions and the electrostatic energy change, which occurs when separating these ions from each other, is much smaller inside water than most of the solvents that are liquid at room temperature and one atmosphere of pressure.

Therefore, a large quantity of ionic liquids and ionic solids, with molecular formula $M_m^{n+}X_n^{m-}$, are subjected to dissolution via a dissociation reaction when mixed in water. At the saturation point the following equilibrium is observed:

$$\underbrace{(M^{n+})_m(X^{m-})_n}_{\text{Ionic bonded}} \rightleftharpoons \underbrace{mM^{n+} + nX^{m-}}_{\text{Solution}} .$$

The maximum solubility of ionic liquids and ionic solids is an important parameter in practical preparative and analytical applications of ESTs. For instance, some analytes are not soluble above certain concentrations, which must be known to avoid errors in the analysis. The same is true for buffers within certain temperature ranges. The maximum solubilities of analytes and buffers at a given temperature are tabulated in the literature and are usually expressed in grams per 100 milliliters of solution.

Another way to express maximum solubility is the so-called *solubility product*, which is defined as $K_s = a_{M^{n+}}^m \times a_{X^{m-}}^n$. This is also well recorded in the literature and is very useful for the theoretical modeling of problems in inorganic chemistry.

The above paragraphs merely give qualitative discussions of the phenomena involved in dissociation reactions. Gibbs free energy change can actually be used to accurately predict if a reaction (dissolution with dissociation) may occur or not (see Section 1.1.8). It is fair to make a clear distinction between dissociation and ionization. The so-called ionic substances (liquids or solids) are predominantly in the form of cations and anions. Therefore, their solubilization in water is merely dissolution with dissociation. On the other hand, a large number of substances are predominantly in the form of covalent liquids, solids, or even gases. Only when placed in water (and a few other solvents) do they ionize to the form of cations and anions. In this case the process is better called dissolution by ionization.

1.1.6 Ionization

Water molecules constantly collide with each other inside pure liquid water. In some collisions a proton of one water molecule attaches to the sp^3 non-bonding orbital of a second water molecule, as given by: $2H_2O \rightarrow H_3O^+ + OH^-$. The generated ions (hydron and hydroxyl) are then immediately solvated and in a fraction of the events the solvated ions move apart as a cation (H_3O^+) and an anion (OH^-), driven by thermal energy and facilitated by the low restoring force, which occurs because of the high relative permittivity ϵ_r of bulk water (equation 1.1). At a given temperature there is a constant rate of ion production and recombination (H_3O^+ + $OH^- \rightarrow 2H_2O$), which leads to an equilibrium denoted by:

$$2H_2O \rightleftharpoons H_3O^+ + OH^-. \tag{1.3}$$

This phenomenon is the so called *self-ionization of water*, i.e. water itself undergoes ionization due to its remarkable properties (*water falls victim to itself*). Moreover, this reaction shows the *amphiprotic* nature of water, as it possesses both the characteristics of a *Brønsted acid* and base.

Molecules carrying acidic and basic functional groups are also subjected to ionization when introduced into water. For carboxylic acids (e.g., formic acid) there are at least two possible reactions for their ionization:

$$HCOOH + H_2O \rightleftharpoons HCOO^- + H_3O^+ \text{ and} \tag{1.4}$$

$$HCOOH + OH^- \rightleftharpoons HCOO^- + H_2O, \tag{1.5}$$

where the source of the OH^- ions is the self-ionization of water. These can be seen as *heterolytic substitution reactions*; note that both H_2O and OH^- work as Brønsted bases in this case. For the ionization of amines (e.g., the secondary amine N-methylmethanamine) there are also at least two possible reactions that can occur:

$$(CH_3)_2NH + H_2O \rightleftharpoons (CH_3)_2NH_2^+ + OH^- \text{ and} \tag{1.6}$$

$$(CH_3)_2NH + H_3O^+ \rightleftharpoons (CH_3)_2NH_2^+ + H_2O. \tag{1.7}$$

Here, both H_2O and the H_3O^+ cation function as Brønsted acids. For a detailed discussion of dissociation and ionization in water see Soustelle (2016).[4]

Finally, it is important to remember that a small fraction of the self-ionization reactions of water produce $H_5O_2^+$ and an even smaller fraction produce $H_7O_3^+$, as well as other ions.

A great deal of knowledge is continuously generated about the internal structure and dynamics of liquid water. Topics such as self-ionization of water, the ions produced, the average life-time of these ions, solvation, and water clusters are difficult to treat because they can not be treated as events in a continuous, homogeneous fluid. They should be treated within the *many-body problem* theories; however, the number of entities in the case of water must be great to be realistic because the water molecules are very sticky with each other due to hydrogen bonding. So one can always remember a saying that remains true: "Of all known liquids, water is probably the most studied and the least understood".[5]

1.1.7 Hydrophilicity, Hydrophobicity, and LogP

Hydrophilicity and hydrophobicity are qualitative terms that refer to chemical substances that, respectively, dissolve in water (strong affinity for water) or in non-polar substances (weak affinity for water). Solvation and the formation of hydrogen bonds are important processes involved in the dissolution of hydrophilic solutes in water (an environment rich in hydrogen bonds). On the other hand, no solvation or hydrogen bond formation occurs when attempting to dissolve hydrophobic (non-polar) substances in water. The dissolution of non-polar substances in non-polar solvents occurs because the positive entropy change of the system (solvent+solute) and the action of van der Waals forces among the hydrophobic solutes and the non-polar solvents. Note that van der Waals forces are much weaker than hydrogen bonds.

Microdrops of hydrophobic substances (e.g., vegetable oil) that are dispersed in water tend to irreversibly join (fuse) when random Brownian motion causes them to collide with each other. Each fusion event causes a sudden increase in the total number of hydrogen bonds in the given volume of water. Macrodrops are formed and at the end expelled from the bulk of the water, which again suddenly increases the total number of hydrogen bonds among water molecules. Note that the inverse reaction (converting a macrodrop of oil into microdroplets) is only possible with the input of a significant amount of energy (usually from vigorous mechanical stirring), as a large quantity of hydrogen bonds must be broken. *Thus, it would be fairer to call water oleophobic instead of calling non-polar substances hydrophobic.*

LogP (or CLOGP) is a well defined quantitative parameter that is related to the hydrophilicity and hydrophobicity of a chemical substance. It is a real number that goes from minus infinity to plus infinity ($-\infty < LogP < \infty$) and is considered a fundamental parameter in ESTs and in many other fields (e.g., organic chemistry and pharmaceutical sciences).

Table 1.2 Examples of some LogP values. These are the average values of the *n* most representative items of each data set. Measurements were taken at room temperature using pure water. The only exception was L-nicotine, which was measured at pH 10.3. (Source: Data from DDBST GmbH, Oldenburg, Germany, www.ddbst.com.).

Analyte	LogP	n
Acetaldehyde	0.43	1
Acrylamide	−0.937	3
Aniline	0.975	16
Benzo[a]pyrene	6.898	11
Caffeine	−0.011	4
Dimethylnitrosamine	−0.570	2
L-nicotine	1.39	1

The LogP of a chemical substance is defined as the decadic logarithm of its 1-octanol/water partition coefficient. The substance is dissolved in water or 1-octanol and then equal volumes of these solvents (one of which contains the substance under analysis) are placed in contact with each other and left to reach equilibrium. When equilibrium is reached the value of *P* is calculated using the following equations:

$$P = \frac{c(\text{in 1-octanol})}{c(\text{in water})} \quad \text{and} \quad \text{LogP} = \text{Log}\left(\frac{c(\text{in 1-octanol})}{c(\text{in water})}\right). \tag{1.8}$$

Table 1.2 gives the LogP values of a few chemical substances. Such values are already used in structure–property correlations and quantitative studies of structure–activity relationships. As more LogP values become available, they will be used in pharmaceutical sciences, pharmanalysis, biochemistry, and analytical chemistry more frequently.

1.1.8 Gibbs Free Energy Change

The solubility (due to dissolution and/or dissociation and/or ionization) of a specific quantity of solute in a certain volume of a given solvent at temperature *T* can be predicted by calculating the Gibbs free energy change: $\Delta G = \Delta H - T\Delta S$. The temperature and changes of enthalpy (ΔH) and entropy (ΔS) all affect the solubility of a solute. The Gibbs free energy change mathematically expresses the role of each of these three thermodynamic variables within reactions (including dissolution). Regardless of the type of reaction (a dissolution, a dissolution with dissociation or a dissolution with ionization), all chemical reactions follow the rule: if $\Delta G < 0$ for a given amount of solute and solvent then the reaction occurs (the solute is soluble), although the velocity may be slow and depends on the potential barriers, but if $\Delta G > 0$ then the reaction will not occur (the specified quantity of solute is not soluble). Generally it is easy to estimate ΔH; however, it is very difficult to evaluate ΔS in most real situations. Dissociation of a salt, for instance, increases the number of states available to the sub-system salt, as solvation occurs, but decreases the number of states available to the sub-system water molecules. This works as a trade off: as the entropy of the salt ions increases the entropy of the water molecules is consequently decreased. When the overall entropy change is negative then lower temperatures favor the occurrence of the dissolution. By contrast, if the entropy change is positive then higher temperatures will favor dissolution. It is necessary to note that lower temperatures generally decrease the velocity of reactions, since a lower thermal energy decreases the probability of reactants overcoming the energy potential barriers of the inter-mediate products. Remember that entropy is proportional to the logarithm of the number of states available to the system, within the constrains (volume and total energy) imposed by the system. A didactic explanation of enthalpy, entropy, and Gibbs free energy is given by Connors (2002) [6] and a quantitatively rigorous approach is described by Reif (1965) [1]. The above presented theory is also valid for the solubilization processes of hydrophobic solutes (and colloids) in aqueous solutions with the aid of micelles (surfactants) or microemulsions (stabilized submicron droplets of oil). The same can be told with the inverse, the solubilization of hydrophilic solutes in hydrophobic solvents.

1.1.9 Acid Ionization Constants

From reaction 1.3, the ionization constants K and K_w of water can be written as:

$$K = \frac{a_{OH^-} \times a_{H_3O^+}}{a_{H_2O} \times a_{H_2O}} \quad \text{or} \quad K_w = a_{OH^-} \times a_{H_3O^+}, \tag{1.9}$$

where the multiplication operation is explicitly denoted by $\times$ to avoid confusion, $K_w = K \times a_{H_2O}^2 \simeq 10^{-14}$ at 25 °C and a_i represents dimensionless variables called chemical activities. The *activity* a_i of a chemical species i is defined as:

$$a_i = \gamma_{c,i}\frac{c_i}{c^{\ominus}} = \gamma_{b,i}\frac{b_i}{b^{\ominus}} = \gamma_{\rho,i}\frac{\rho_i}{\rho^{\ominus}}, \tag{1.10}$$

where γ is a dimensionless parameter called the *activity coefficient*, which depends on the units of concentration of the variables $c^{\ominus}$, $b^{\ominus}$, and $\rho^{\ominus}$. These are *standard states* of solute concentrations with the following units, respectively: amount concentration (molar), molality (molal), and mass concentration (g L^{-1}). They should not be confused with the *standard solutions* used in analytical chemistry, nor with the *standard conditions* of a system (e.g., standard temperature and pressure of a gas). These standard states are standard quantities of a thermodynamic variable and in the present case could be 1 M, 1 molal, and 1 g L^{-1}.

The activity coefficients express the deviation from an ideal behavior. When the activity coefficient γ_i of a chemical species i is close to one for a given range of concentration amount or other unit, then this species exhibits an almost ideal behavior according to *Henry's law* in this range and the same is expected up to infinite dilutions of the solute.

The equilibrium constant K_w is called the *autoprotolysis constant*, [7] the water dissociation constant, the ionization constant or self-ionization constant of water. From the definition shown in equation 1.9 it may also be seen as the *ionic product of water*. These are small numbers that are difficult to handle. Therefore it is more practical to apply the mathematical operator "p", which stands for "$-\log$", to them. Consequently we obtain:

$$pK_w = p_{OH^-} + p_{H_3O^+}.$$

In reality it is more common to use the simplified notations of pH and pOH instead of $p_{H_3O^+}$ and p_{OH^-}, respectively. For all other entities the notation p_x, where x denotes any charged or neutral species, is used. For example, p_{Cl^-}, p_{Na^+}, p_{HCOO^-}, $p_{NH_4^+}$, ..., and so on.

From the mathematical point of view it is incorrect to write pH $= \log c_{H^+}$ or pH $= \log [H^+]$, because the transcendental functions (exponential, logarithmic, and trigonometric) must be handled with dimensionless arguments. Second, from a chemical point of view, the cited expressions are not very informative. To illustrate this, let us suppose that the pH of a solution is exactly pH $= -\log a_{H_3O^+} = 2$. From equation 1.10 we can see that this is much more information rich, as it leads to the following relationships:

$$c_{H_3O^+} = \frac{c_{H_3O^+}^{\ominus}}{\gamma_{c,H_3O^+}} 10^{-2}, \qquad (1.11)$$

$$b_{H_3O^+} = \frac{b_{H_3O^+}^{\ominus}}{\gamma_{b,H_3O^+}} 10^{-2}, \quad \text{or} \qquad (1.12)$$

$$\rho_{H_3O^+} = \frac{\rho_{H_3O^+}^{\ominus}}{\gamma_{\rho,H_3O^+}} 10^{-2}. \qquad (1.13)$$

These allow the content of H_3O^+ to be known in many more units (including, but not limited to, molar, molal, and g L^{-1}) and with much higher precision as more parameters of $\gamma_{x,i}$ (with $x = c, b, \rho, ...$) become available.[8]

Small molecules carrying an acid and/or a basic group tend to be soluble in pure water and exhibit a certain degree of ionization (Section 1.1.6) or dissociation (Section 1.1.5). The ionization or dissociation of the acid HA in aqueous solutions

$(H_2O + HA \rightleftharpoons A^- + H_3O^+)$ at a given temperature is characterized by the acid ionization constants K and K_a. They are defined as:

$$K = \frac{a_{A^-} \times a_{H_3O^+}}{a_{H_2O} \times a_{AH}} \quad \text{and} \quad K_a = \frac{a_{A^-} \times a_{H_3O^+}}{a_{AH}}, \tag{1.14}$$

where a_i is the activity of the species i. The activity of water is approximately constant at a given temperature and with the low solute concentrations normally used in the ESTs. Therefore, it is assumed that $K_a = K \times a_{H_2O}$. Equivalently, the conjugated acid BH^+ of base B, produced by the equilibrium reaction $H_2O + BH^+ \rightleftharpoons B + H_3O^+$, has the following acid ionization constants:

$$K = \frac{a_B \times a_{H_3O^+}}{a_{H_2O} \times a_{BH^+}} \quad \text{and} \quad K_a = \frac{a_B \times a_{H_3O^+}}{a_{BH^+}}. \tag{1.15}$$

These acid ionization constants are better represented by pK_a, where $pK_a = -\log K_a$.

The pK_a ranges defined for very strong ($pK_a < 0$), strong ($pK_a \sim 0$), medium, weak, and very weak acids are poorly defined. Nevertheless, it is safe to say that acids with a pK_a of between 4 and 10 are weak acids and that acids with $pK_a > 10$ can be considered as very weak acids. The opposite occurs for bases, as for a base to be very strong the pK_a of the conjugated acid BH^+ (equation 1.15) must be above 14 ($pK_a > 14$), with strong bases exhibiting $pK_a \sim 14$. Similar to weak acids, weak bases also exhibit pK_a of between 4 and 10; however, in this case the very weak bases exhibit $pK_a < 4$.

It is important to note that pK_a significantly changes with temperature for some functional groups: $pK_a = f(T)$, where f is expected to be a smooth and slowly varying function of temperature. The pK_a of the great majority of amines decreases with temperature, while carboxylic acids exhibit a much smaller change, usually negative, but there are some exceptions and it depends on the temperature range. These temperature sensitivities have important practical implications for method development within the field of ESTs, as they affect the mobility of the analytes and the pH of the buffers.[9–11] Moreover, they are used to promote cyclic band compression in the toroidal layouts, which is an interesting way to get some control of band spreading along the separation mediums (see Appendix G).

1.1.10 Concentration–pH and pa–pH Diagrams

It is important to know the concentration of all chemical species present in a given buffer solution at a given pH, because some species may interfere with the migrating analytes under study. Additionally, it is also important to know the concentrations of all existing species of an analyte present in a separation medium at a given buffer pH. This is important because the analyte species present define the average electrophoretic mobility of the analyte, the system peaks, and the interactions of the species with the BGE components. c–pH diagrams are one of

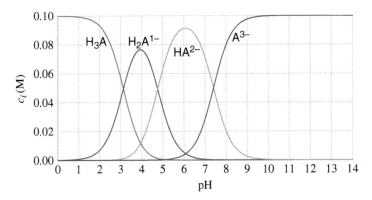

Figure 1.4 c–pH diagram of citric acid species in an aqueous solution in the 0 to 14 pH range. 0.1 mol of citric acid was used to prepare 1 L of solution, i.e. $c^{\circ}_{H_3A} = 0.1$ M.

the most commonly used tools to visualize the concentration of chemical species, showing the concentration of each species at every pH in the $0 < pH < 14$ range (which is the maximum range used in ESTs). Figure 1.4 shows the c–pH diagram of a triprotic acid (0.1 M citric acid in an aqueous solution), which is a complex acid because its successive ionizations produce many species. In Figure 1.4 only the most concentrated species can be seen, as the curves of the most diluted species (with $c < 0.001$ M) run too close to the line at $c = 0$ M to be observed. Nevertheless, many detectors (fluorescence and potentiometric) are able to detect species down to 10^{-10} M and even lower concentrations. Therefore, sometimes it is desirable to visualize the concentration profiles of species within lower concentration ranges. pa–pH diagrams are ideal for this as they show pa_i as a function of pH. Figure 1.5 shows the same case as studied in Figure 1.4 (c–pH diagram of citric acid at the initial concentration of $c^{\circ}_{H_3A} = 0.1$ M), but it is now represented as a pa–pH diagram.

Figures 1.4 and 1.5 are nothing more than semi-log and log-log plots, respectively. Figure 1.4 allows the visualization of concentration changes of the main species at $pH = pK_{a1}$ (3.13), pK_{a2} (4.76), and pK_{a3} (6.39). On the other hand, Figure 1.5 allows the visualization of the concentrations of extremely diluted species.

1.1.11 Henderson–Hasselbalch Equation

Rearranging the terms on the right hand side of equations 1.14 and 1.15, as well as applying the "$- \log$" operator to them, produces the following:

$$pH = pK_a + \log \frac{a_{A^-}}{a_{HA}} \quad \text{and} \quad pH = pK_a + \log \frac{a_B}{a_{BH^+}}. \tag{1.16}$$

(a) Weak acid + strong base   (b) Weak base + strong acid

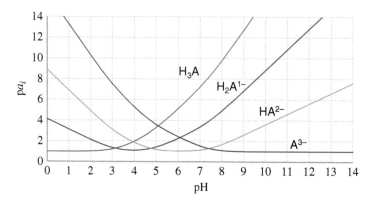

Figure 1.5 pa–pH diagram of citric acid species in an aqueous solution. The conditions are identical to those used in Figure 1.4. Here, however, the concentration of the species can be seen for broader ranges.

Within diluted solutions, the activities a_i can be replaced by $a_i = \gamma_i^c\, c_i/c_i^\ominus$ to give $a_i/a_j \approx c_i/c_j$. Therefore:

$$\mathrm{pH} = \mathrm{p}K_a + \log \frac{c_{A^-}}{c_{HA}} \quad \text{and} \quad \mathrm{pH} = \mathrm{p}K_a + \log \frac{c_B}{c_{BH^+}}. \tag{1.17}$$

These equations are very useful for the preparation of buffer solutions with a given pH. One way to prepare a buffer solution is by adding a strong base to a solution that contains a weak acid dissolved in it. The concentration of c_{A^-} will be negligible if compared to c_{HA}° (amount of initial weak acid) before the addition of the strong base. This comes from the definition of a weak acid. Therefore, the addition of x mol of a strong base to a solution containing y mol of a weak acid will result in x mol of A^- ions, and $y - x$ mol of HA molecules will be left in the solution. The resulting pH will then be given by: $\mathrm{pH} = \mathrm{p}K_a + \log \frac{x}{y-x}$, with the condition that $0 < x < y$. On the other hand, if the weak acid is not so weak (or the weak base is not so weak) then the final pH will be slightly lower (or higher) than predicted. This occurs because extra A^- ions, and consequently some H^+ ions, come from the self-ionization of the acid itself and not just from the added strong base (or strong acid).

Besides precise pHs values, the buffer capacity of a given buffer solution is also a very important parameter as it indicates how good the buffer is at keeping the pH nearly constant throughout the addition of a small quantity of a strong acid or base. Section 1.1.12 shows that the highest buffer capacities of monoprotic acids are observed for pH values close to $\mathrm{p}K_a$. The maximum buffer capacity of a multiprotic acid can be visualized by plotting its buffer capacity against pH (also shown in

Section 1.1.12). Finally, equation 1.17 can be arranged into the more useful format:

$$x = \underbrace{\frac{y}{1 + 10^{pK_a - pH}}}_{\text{Weak acid+strong base}} \quad \text{and} \quad z = \underbrace{\frac{w}{1 + 10^{pH - pK_a}}}_{\text{Weak base+strong acid}}, \tag{1.18}$$

where $x(z)$ are the number of moles of a strong base (strong acid) that must be added to a solution containing $y(w)$ moles of a weak acid (weak base), characterized by pK_a, to obtain a final solution with the desired pH. This can then be diluted until the desired concentration is achieved. The ideal situation is to have: $(pK_a - 0.5) < pH < (pK_a + 0.5)$, as shown in Section 1.1.12.

Equation 1.18 will still be valid for multiprotic acids or molecules bearing a mixture of acidic and basic functional groups (zwitterionic compounds or zwitterions) if their pK_a values are separated from each other by at least two units. Examples of such molecules include most proteinogenic amino acids and linear peptides without ionizable lateral chain residues. Within some pH ranges, however, if the pK_a values are less than two units apart then two or more anions (or cations) from the same acid (or base) will start to coexist at concentration of more than 1% of the total molar concentration of the initial acid (or base). In this case the idea of separately visualizing conjugated acids and bases becomes a little bit confusing as the same species may start to play both roles. This is well demonstrated within the c–p and p–p diagrams. Luckily, buffer capacity (see Section 1.1.12) can still be calculated and measured under these circumstances (pK_a values, which are less than two units apart from each other).

1.1.12 Buffer Capacity

The buffer capacity (β) of a buffer solution is a measurable quantity and is defined as:

$$\underbrace{\beta = \frac{dn}{d(pH)}}_{\text{(a) Differential form}} \quad \text{or} \quad \underbrace{\Delta n = \int_{pH_1}^{pH_2} \beta \, d(pH).}_{\text{(b) Integral form}} \tag{1.19}$$

β is defined as the derivative of the amount of equivalents (of a strong acid or base) to the pH. The unit of β is the amount of equivalents that must be added to one liter of a buffer solution in order to change pH by one unit. Buffer solutions with high buffer capacities are always desired. In the laboratory β is measured by titration and observing the amount of equivalents (δn_1) of a strong base (or acid) that must be added to one liter of a buffer solution with an initial pH of pH_1, in order to increase (or decrease) its pH from pH_1 to pH_2. This permits the definition of $\delta pH_1 = pH_2 - pH_1 \ll 0.1$ (depending on the sensitivity and precision of the pH-meter used). At this point the titration is paused and the buffer capacity of the solution at pH_1 is calculated as follows: $\beta(pH_1) = \delta n_1 / \delta pH_1$. Following

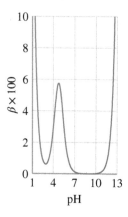

Figure 1.6 Buffer capacity of 0.1 M acetic acid ($pK_a = 4.75$) in an aqueous solution in the 1 to 13 pH range.

this the titration is continued until the pH is changed by another increment of $\delta pH_2 = pH_3 - pH_2$. $\beta(pH_2)$ is then calculated as $\beta(pH_2) = \delta n_2 / \delta pH_2$, and so on. Plotting the β values of a buffer against pH gives nothing more than a density probability distribution.

Equation 1.19(b), which is the integral form of equation 1.19(a), shows the number of equivalents of a strong base (or a strong acid) required to change the pH of one liter of a buffer solution from pH_1 to pH_2 (or the reverse). This integral represents the area under the curves $\beta(pH)$ between pH_1 and pH_2.

For practical applications, buffer capacity can be predicted using equation 1.19 (a) and the definition of acid ionization constant (equation 1.15). The result is the following equation:

$$\beta = \ln(10) \left(\frac{K_w}{10^{-pH}} + 10^{-pH} + \frac{a\,K_a\,10^{-pH}}{(K_a + 10^{-pH})^2} \right), \tag{1.20}$$

where $K_w = 10^{-14}$ (at 25 °C), $K_a = 10^{-pK_a}$, and a is the initial activity of the weak acid used to prepare the buffer solution. For instance, if 0.1 mol of acetic acid is used to prepare one liter then $a = \gamma_c\, c^\circ / c^\ominus = 0.1M/1M = 0.1$ if $\gamma_c = 1$. The same equation can be applied to a weak base (B), however in this case a represents the initial activity of the weak base used.

Figure 1.6 shows the buffer capacity that should be expected when 0.1 mol of acetic acid is diluted in water to make 1 L solution. This gives a 0.1 M solution and note the high buffer capacity around pH = 4.75, which is the pK_a of acetic acid. Figure 1.7 shows the same for a 0.1 M solution of Tris.

If a buffer is prepared by a mixture of n weak acids and a strong base (or a mixture of n weak bases and a strong acid), then the buffer capacity can still be predicted for each pH as follows:

$$\beta = \ln(10) \left(\frac{K_w}{10^{-pH}} + 10^{-pH} + \sum_{i=1}^{n} \frac{a_i\,K_{a,i}\,10^{-pH}}{(K_{a,i} + 10^{-pH})^2} \right). \tag{1.21}$$

Figure 1.7 Buffer capacity of 0.1 M Tris (pK_a = 8.07) in an aqueous solution.

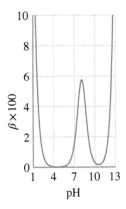

Figure 1.8 Buffer capacity of 0.1 M β-alanine (pK_a = 3.63 and 9.6) in an aqueous solution. The buffer capacity at the isoelectric point (pH(I) = 6.6) is low.

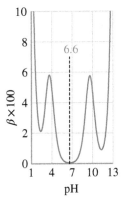

In this case the index i refers to each weak acid (or weak base) or each functional group if the same molecule carries weak acid groups and weak basic groups. Figures 1.8 and 1.9 were plotted using this equation.

Figure 1.8 shows the buffer capacity of a 0.1 M solution of β-alanine as a function of pH and Figure 1.9 for a 0.1 M glycyl-aspartic acid. Note that both β-alanine and glycyl-aspartic acid do have isoelectric points (IEP), however only glycyl-aspartic acid has a good buffer capacity around its or close to the pH(I). A high buffer capacity at or close to the pH(I) is a highly desired property in the ESTs, as it allows the pH to be stabilized with minimal increase in the conductivity.

The total conductivity of the buffer must be higher than or close to the conductivity of the sample, otherwise it causes band destacking (Section 2.14). On the other hand, the lower the conductivity of the buffering electrolyte to the total conductivity of the buffer solution the better, as some room is left for the addition of some electrolytes with the same mobility as the analytes. This is important for the

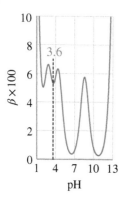

Figure 1.9 Buffer capacity of 0.1 M glycyl-aspartic acid (pK_a = 2.81, 4.45 and 8.6) in an aqueous solution. Note the high buffer capacity at pH(I) = 3.6.

production of symmetric peaks in most of the separation modes. Isotachophoresis and isoelectric focusing are exceptions to this rule as they are based on special separation mechanisms. Some samples do have low conductivities, in this case the development of buffer solutions with low conductivity, i.e. designed for high field strengths operations, is always an active topic of research, [12] as they lead to high separation efficiencies (Section 3.5).

Another very useful concept is the *normalized buffering capacity/conductivity ratio*.[13] Here, the buffering capacity at each pH is divided by the buffer conductivity at this pH. A high buffer capacity and low buffer conductivity are highly desirable. This allows different buffers to be compared among themselves at different pHs. Such performance rankings of good buffers are still missing in the literature.

1.2 Binary Mixtures and Other Solvents

Besides pure water, some binary mixtures of water and a protogenic organic polar solvent (methanol, ethanol or 2-propanol) or a protophilic organic polar solvent (acetonitrile, acetone or dimethylsulfoxide) have been studied and applied as solvents in ESTs [14,15], over the years. For instance, water-methanol, [16] water-ethanol, [17] water-isopropanol, [18] and water-acetonitrile. [19] For some applications they are necessary for many reasons. Firstly, to improve the solubility of analytes that exhibit poor solubility in pure water. Secondly, to control the electroosmotic flow in order to shorten or extend the run time. Apparent mobility (from injection point to detection point) can be increased for analytes that are easily separated, shortening the run time. This decreases the distance run by the analytes in the reference frame of the liquid phase, providing the benefit of a shorter run time but consequently decreasing the separation efficiency. Many mixtures are difficult to separate and in this case the electroosmotic flow is set to run in the opposite direction to the analytes. This means that the analytes now run longer distances in the reference frame of the liquid, which means that the separation efficiency is tremendously increased. However, as discussed in many Sections (e.g. 2.9 and 2.10), electroosmosis is difficult to control and exhibits poor repeatability from run to run. This has a negative impact on peak identification and quantitative analyses. Third, a number of other minor benefits occur depending on the amount of organic modifier present; these benefits include: many binary mixtures exhibit a much longer shelf life (microbial growth is inhibited); some can be stored at $-8\,^{\circ}$C without freezing, so that they reach working temperature much quicker than frozen solutions; standard solutions of biomolecules last much longer in these mixtures; and the dynamic viscosity of the mixture can be increased/decreased, which is beneficial when working with micro-channels and capillaries with a broad range of diameters.

Finally, there are a few applications (for instance the separation of aminium ions, quaternary ammonium ions, and coordination complexes) that can be run with pure propylene carbonate, formamide or other solvents, since the analytes remain charged within them. Moreover, propylene carbonate, for instance, does not produce gases at the electrodes, allowing the separation to be run in sealed systems.

References

1 Reif, F. (1965). *Fundamentals of Statistical and Thermal Physics*. Tokyo: McGraw-Hill Kogakusha.

2 Rahm, M., Zeng, T., and Hoffmann, R. (2019). *Journal of the American Chemical Society* 141: 342–351.

3 Wohlfart, C. and Lechner, M.D. (eds.) (2008). *Static Dielectric Constants of Pure Liquids and Binary Liquid Mixtures*, Landolt–Börnstein. Berlin: Springer-Verlag.

4 Soustelle, M. (2016). *Ionic and Electrochemical Equilibria*. Hoboken, NJ: Wiley.

5 Franks, F. (1979). *Water: A Comprehensive Treatise: Recent Advances*, vol. 6. Boston, MA: Springer-Verlag.

6 Connors, K.A. (2002). *Thermodynamics of Pharmaceutical Systems*. New York: Wiley.

7 IUPAC (2012). Compendium of Chemical Terminology, Gold Book, Version 2.3.2. International Union of Pure and Applied Chemistry.

8 Pitzer, K.S. (ed.) (1991). *Activity Coefficients in Electrolyte Solutions*, 2e. Boca Raton, FL: CRC Press.

9 Reijenga, J.C., Gagliardi, L.G., and Kenndler, E. (2007). *Journal of Chromatography A* 1155: 142–145.

10 Reijenga, J.C. (2009). *Journal of Chromatography A* 1216: 3642–3645.

11 Gagliardi, L.G., Tascon, M., and Castells, C.B. (2015). *Analytica Chimica Acta* 889: 35–37.

12 Hjertén, S., Valtcheva, L., Elenbring, K., and Liao, J.L. (1995). *Electrophoresis* 16: 584–594.

13 Stoyanov, A.V. and Righetti, P.G. (1998). *Electrophoresis.* 19: 1674–1676.

14 Schwer, C. and Kenndler, E. (1991). *Analytical Chemistry* 63: 1801–1807.

15 Sarmini, K. and Kenndler, E. (1997). *Journal of Chromatography A* 792: 3–11.

16 Sarmini, K. and Kenndler, E. (1998). *Journal of Chromatography A* 606: 325–335.

17 Sarmini, K. and Kenndler, E. (1998). *Journal of Chromatography A* 811: 201–209.

18 Sarmini, K. and Kenndler, E. (1998). *Journal of Chromatography A* 818: 209–215.

19 Sarmini, K. and Kenndler, E. (1999). *Journal of Chromatography A* 833: 245–259.

2

Fundamentals of Electrophoresis

2.1 Introduction

In this chapter, the fundamentals of electrokinetic separation techniques (ESTs) are presented, so that the open and toroidal layouts can be easily compared throughout this book. Electrophoresis and electroosmosis, which are the two most important *electrokinetic phenomena* [1] used in ESTs, are presented in detail. The other examined topics include the definitions of ionic limiting mobility, bands, fronts, peaks, zones, isoelectric points, turbulent flow, laminar flow (in cylindrical and rectangular conduits), the Joule effect, heat dissipation, temperature profiles, and band broadening mechanisms. The difference between sample stacking and band compression is also briefly examined. Finally, the nine most widely used separation modes are presented.

2.2 The Platforms

Electrophoresis platforms are the physical structures on which electrokinetic separations are performed. They may be microscopic conduits, such as microtubes and microchannels filled with liquid solutions, or macroscopic slabs, which are made from gels (among other materials) and are laid inside electrophoresis chambers.

Figure 2.1 shows the Cartesian coordinates (x, y, z) adopted in this book for microchannels and slabs. The x direction always denotes the direction along which the electric potential gradients are applied and the separations are performed. The width of the separation medium is denoted by $2w$, and consequently $-w < y < w$. Additionally, the height is denoted by $2h$, and therefore $-h < z < h$. This notation has some advantages, as shown in Section 2.8.4 and Appendices C and D.

Figure 2.2 shows the cylindrical coordinated axis used throughout this book (r, θ, x). As commonly assumed, x denotes the axis along which electrophoresis separations are performed. The r spatial coordinate represents the radial direction,

Open and Toroidal Electrophoresis: Ultra-High Separation Efficiencies in Capillaries, Microchips, and Slabs, First Edition. Tarso B. Ledur Kist.
© 2020 John Wiley & Sons Ltd. Published 2020 by John Wiley & Sons Ltd.

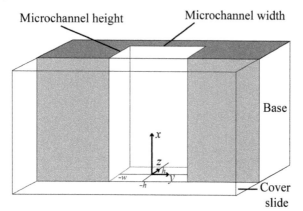

Microchannel height   Microchannel width

Base

Figure 2.1 The Cartesian coordinates (x, y, z) used in this book for the microchannels in microchip. Similar axis orientation is also applicable for square fused silica microtubes (square capillaries) and slabs.

Cover slide

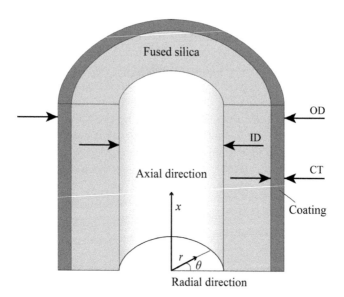

Fused silica

OD

ID

Axial direction

CT

Coating

Radial direction

Figure 2.2 The cylindrical coordinates (r, θ, x) used in microtubes and flexible fused silica microtubes (capillaries). CT stands for polymer coating, ID for inner diameter, and OD for outer diameter. The inner radius is denoted by a ($a = \text{ID}/2$).

which runs orthogonal to the capillary wall. The values of r in the liquid core go from zero to a, which is the internal radius or half the internal diameter (ID) of the capillary. The angle θ goes from zero to 2π and, together with r, defines the points of a plane which is orthogonal to the axial direction x. The cylindrical micro-tubes or capillaries are usually made of fused silica. 10 to 30 μm of an external polymer coating (CT), as shown in Figure 2.3, prevents them from breaking and instead makes them flexible. In the open and toroidal layouts, the capillaries are bent

Figure 2.3 Illustration of three flexible fused-silica microtubes (capillaries) showing the coating thickness (CT), inner diameter (ID), and outer diameter (OD).

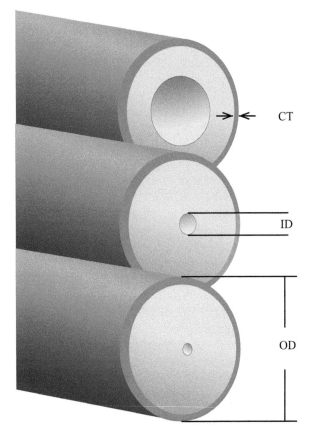

CT

ID

OD

and even coiled for use. Even so, the cylindrical coordinates still represent a good approximation of straight capillaries as the commonly used lengths L are always much larger than a. A short capillary with $L = 20$ cm and a large ID of 100 μm, produces $L/a = 4000$.

2.3 Electrophoresis

The first systematic experiments with Electrophoresis were reported by Faraday back in 1791, when he presented the laws of electrolysis. Later, von Helmholtz (1877), Hittorf (1858), Arrhenius (1887), Nernst (1897), Kohlrausch (1897), and Tiselius (1930) made fundamental contributions to the field. These historical milestones are well reviewed by Vesterberg.[2,3]

Electric currents are observed when electric potential gradients are applied to liquids containing electrolytes. They are applied with the aid of two electrodes made of an inert conductive material, which are immersed into solutions containing electrolytes. According to the *theory of electromagnetism* [4], in the one-dimensional case the applied electric field $E(x, t)$ is related to the applied electrostatic potential gradient according to the following relation:

$$E(x,t) = -\underbrace{\frac{\partial V(x,t)}{\partial x}}_{\text{If } \partial V/\partial x \text{ changes over space}} \quad \text{or} \quad E(t) = -\underbrace{\frac{V(L,t) - V(0,t)}{L} = -\frac{V(t)}{L}}_{\text{If } \partial V/\partial x \text{ is constant from } x = 0 \text{ to } L}, \qquad (2.1)$$

where L represents the distance over which the electric potential drops by the amount of $V(t) = V(L,t) - V(0,t)$. In the great majority of ESTs these applied potentials are constant in time during the runs and also have a constant spatial gradient $(\mathrm{d}V/\mathrm{d}x)$ from $x = 0$ to $x = L$ (one important exception to this is isotachophoresis, see Section 2.15.6). In this case the electric field strength is constant and given by $E = V/L$, where V is the applied electrostatic potential difference usually maintained by a regulated high voltage source or module. The typical values used range from 100 V to 500 V for slabs, 500 V to 5000 V for microchips and 5000 to 30 000 V for capillaries.

According to the theory of electromagnetism an electric field E pulls ions. As cations migrate to the cathode (electrode with a negative or lower potential) and anions migrate to the anode (electrode with a positive or higher potential), their movement produces an electric current (also called ionic current). These ions are the charge carriers; anions bear electrons and cations bear positive charges (or a lack of electrons). This is the basic description for conductance of electrolytes in solutions. The phenomenon underlying conductance is called electrophoresis (phoresis or phoresy originate from the Greek word *phoras*, meaning bearing).

Electrophoresis is characterized by *electrophoretic mobility* (μ_e), which is defined as: the average observed rate of migration or velocity (v) of a molecule or particle in the reference frame of the liquid, divided by the applied electric field strength (E), i.e. $\mu_e = v_e/E$. A mathematical expression for μ_e can be deduced and does not depend on the layout used (open or toroidal) or on the platform where it is performed (capillary, microchip or slab). It only depends on the friction coefficient and charge of the ion, molecule, or particle; meaning that only the electric and friction forces are acting on it. In this case $\mu_e = q/f$, where q is the charge and f the viscous friction coefficient. This can be calculated from first principles using Langevin's equation, as shown in Appendix C.

Electrical mobility (μ_{el}), on the other hand, is a more general term. This transport phenomena is responsible for the movement of both charged and uncharged entities. It also does not depend on the layout and platform used and is also calculated as the migration rate (with respect to the buffer solution) divided by the electric field strength, but regardless of the ligands, inclusion compounds, micelles, microemulsions, or any pseudophase that are present. For instance, the electrical mobility of a neutral compound in an aqueous solution may be non-null and depends on the charge and concentration of micelles. This is different from the effective mobility, which is given by: $\mu_{ef} = \alpha\mu_e$, where α is the degree of ionization of a molecule (discussed in Section 2.4).

The additives used and the underlying separation mechanisms defines the so-called separation mode. What differentiates the ESTs is the layout, the platform, and the separation mode. The mathematical expressions of μ_{el} are calculated for some separation modes in Section 2.15. In the "free-solution electrophoresis separation mode" the analytes are separated by the electrophoretic mobility (μ_e) differences only. The free-solution electrophoretic velocities in the steady state regime are given by the following equations:

$$\underbrace{v_e = \mu_e E = \frac{qE}{f} = \frac{qV}{fL}}_{\text{(a) General}} \text{ and } \underbrace{v_e = \frac{qE}{6\pi r \eta} = \frac{qV}{6\pi r \eta L}}_{\text{(b) Spherical entity}}, \tag{2.2}$$

where q is the charge of the ion, molecule, or particle, f is the viscous friction coefficient, which is dependent on the size, shape, and viscosity (η) of the liquid within which electrophoresis takes place, and L is the length of the separation medium over which the applied electric field E is constant and the electrostatic potential difference drops by an amount of V. The right hand side of equation 2.2 is valid only for spherical ions, molecules or particles with a radius of r. Equation 2.2 is deduced in Appendix C (equations C.1–C.5). As shown in Appendix C, v_e and $x = v_e t$ refer to the center of mass of the band/peak and not to the maximum of the bands/peaks.

The expression of the steady state electrophoretic mobility (μ_e), and the distinction between the electrical mobility and the electrophoretic mobility has been discussed in the previous paragraphs. However, even with constant electric fields (E) it must be noted that there are two regimes for these mobilities: the transient regime and the steady state regime. In the steady state regime, discussed in the previous paragraph, the center of mass of the band migrates at a constant average velocity. The transient regime begins when the high voltage is turned on; it is characterized by a velocity change from zero to the steady state velocity. The transient regime is also calculated in Appendix C for free-solution electrophoresis. In this case, if a high voltage is turned on instantly at time $t = 0$, these two regimes are given by:

$$\underbrace{v_e = \mu_e(1 - e^{-t/\tau_e})\, E}_{\text{Transient regime}} \text{ and } \underbrace{v_e = \mu_e E}_{\text{Steady regime}}, \tag{2.3}$$

where τ_e is a real and positive constant given by $\tau_e = m/f$, and m is the mass of the ion or charged molecule or particle. The parameter τ_e is called a time constant and gives the time scale of this event.

Taking as an example the sodium ion (Na$^+$), which has an electrophoretic mobility of 5.19×10^{-8} m^2 V^{-1} s^{-1} in pure water, charge +1, and considering the total mass of the sodium ion plus five water molecules of the inner solvation shell, results in: $\tau_e \sim 10^{-10}$ s. This is a very short transient time. However, for viral particles and cells it may take microseconds or more.

In ultra-pure liquid water there is still a small and constant observable ionic current when an electrostatic potential difference is applied. This *conductivity* is caused by the H$_3$O$^+$ and OH$^-$ ions, having both a concentration of 10^{-7}M $= 0.1$ μM at neutral pH (at pH $= 7$). This ions are always present due to the self-ionization of water (equation 1.3). When more substances are added to pure water, its conductivity will increase if the concentration of the electrolytes increases. This increase of electrolytes can occur either through the dissociation (Section 1.1.5) and/or the ionization (Section 1.1.6) of the added substances. Therefore, the conductivity (or specific conductance) of a solution is related to the concentration of the electrolytes present within it and to the electrophoretic mobility of each electrolyte.

For the electrophoretic mobility the relationship between conductivity (κ) and mobility (μ_e) is given by:

$$\kappa = A \sum_i c_i |\mu_{e,i}| |z_i| F, \tag{2.4}$$

where A is the cross sectional area, F is the Faraday constant, c_i is the concentration, and z_i is the valence of each ionic species.

Note that the reciprocal of conductivity (κ) is called resistivity (ρ), with $\kappa = 1/\rho = A/RL$, where R is the electrical resistance, L the medium length, and A the cross sectional area. The SI unit of conductivity is Siemens per meter (S m^{-1}) and the SI unit of resistivity is Ohms $\times$ meter (Ω m).

At the surface of the electrodes (made out of carbon, graphite, metal, or a conductive polymer) the ionic current is converted into an electric current as the electrons jump (or tunnel) from the arriving anions to the electrodes (at the anode) and from the electrodes to the arriving cations (at the cathode). In order to avoid confusion, the terms positive electrode, where the higher potential is applied, and negative electrode, were the lower potential is applied, can be used.

These induced ionic currents usually inflict some chemical changes on the electrolytes and/or solvents, as some chemical reactions may take place at the electrodes. The cumulative effect of these reactions over time changes the chemical composition of the solutions. These reactions are dependent on the solvent used and the nature of the electrolytes. In aqueous solutions it is common to see gas bubbling on the surface of the electrodes, which is called water electrolysis if O$_2$ is generated at the positive electrode and H$_2$ is generated at the negative electrode. In this case it comes from the following net reactions: $4\text{OH}^- \longrightarrow 2\text{H}_2\text{O} + \text{O}_2{\uparrow} + 4e$ at the positive electrode and $2\text{H}_3\text{O}^+ + 2e \longrightarrow 2\text{H}_2\text{O} + \text{H}_2{\uparrow}$ at the negative electrode.

2.4 Electrophoresis of Single Molecules

There are only a few ways for small molecules to became charged in aqueous solutions within the pH range of 1 to 13. pH values of pH < 1 and pH > 13 are rarely used in the ESTs because they are only achievable by incurring the cost of higher ionic strengths and, consequently, higher conductivity values in most of the cases. These are undesirable as they limit the maximum electric field strength that can be applied along the separation medium. The simplest way to charge a molecule that contains at least one acidic or basic functional group is to adjust the pH of the solution. Consider a molecule with only one acidic functional group (the monoprotic acid HA). One molecular scenario consists of a single aqueous solution, at a given pH, that contains a single probe molecule. In this case, what is the charge of this single molecule at different instants of time (t_i) that are randomly selected between t_1 and $t_1 + \Delta t$? Another scenario consists of an ensemble of solutions at a fixed pH, each containing a single probe molecule. What is the average charge of all these molecules at a given instant in time? These two situations are addressed in the following paragraphs.

Interesting models are used in *molecular dynamic simulations* and the results of these simulations show that at some moments in time (t_i) the charge of the probe molecule will be zero when the molecule is in its HA form. At other points in time the charge will be $-e$ when the molecule will be in its A^- form. Within this book the shielding effect of the nearby ions and solvation shell on the effective charge of these molecules are omitted. This subsection focuses on calculating the time and ensemble average charge considering the two above mentioned states.

The classical way to calculate the mean charge $\langle q \rangle$ of a single molecule is to take a time average. This is achieved by calculating the integral of the molecular charge over time and dividing it by the interval Δt as follows:

$$\langle q \rangle_{\Delta t} = \frac{1}{\Delta t} \int_{t_1}^{t_1+\Delta t} q(t)\, dt \quad \text{with} \quad \Delta t \gg \tau, \tag{2.5}$$

where τ is the mean lifetime of the A^- ion. In this case the function $q(t)$ jumps from zero to -1 and from -1 to zero, as shown in Figure 2.4. The average width of these random rectangles equals the mean lifetime (τ) of the A^- ion. The result will be the sum of all of these areas (between t_1 and $t_1 + \Delta t$) divided by Δt.

There is a second more intuitive way to estimate the mean charge ($\langle q(t) \rangle$) of a single molecule over time. This method is based on sampling random, discrete points of time between t_1 and $t_1 + \Delta t$ and taking note of the observed charges. The average charge is then calculated as follows:

$$\langle q \rangle_{t_i} = \frac{1}{n} \sum_{i=1}^{n} q(t_1 + r_i\, \Delta t) \quad \text{with} \quad 0 \le r_i \le 1, \tag{2.6}$$

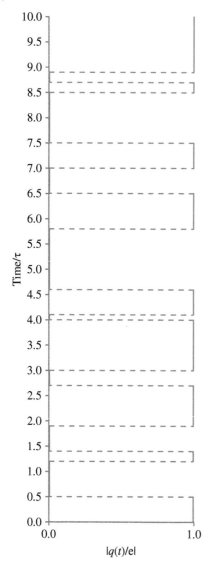

Figure 2.4 Ionizations and recombinations of a molecule along time, represented by jumps (dashed lines) between its neutral (HA) and charged (A⁻) states. Time is given in units of the mean lifetime of the excited state.

where r_i ($i = 1, 2, 3, ..., n$ and $n \gg 1$) is a set of random numbers with a constant probability density distribution between 0 and 1. The addends of this summation consist only of values of 0 and −1, assuming that the jump transition time is negligible compared to τ, and therefore the output of the summation will lie between 0 and −1. These measurements can be made, for instance, by monitoring a single fluorescent molecule, which produces distinct emission bands from its HA and

A^- forms, with two single photon counting detectors (one for each spectral range). The work of Shera et al. [5] was a landmark publication for the measurement of single molecules in liquid solutions using fluorescence at room temperature as it was the first publication of its kind.

The calculation shown by equation 2.6 is, coincidentally, the same as the *Monte Carlo integration technique* that is used in *numerical calculus* to solve definite integrals like the integral of equation 2.5 (area of Figure 2.4). This is based on the use of random numbers to solve definite integrals. Equation 2.6 and the Monte Carlo method are identical by coincidence only, which is an indication of the consistency of the above reasoning. In other words, two distinct mathematical methods of calculating definite integrals were proved equivalent using physico-chemical arguments.

There is even a third way to calculate the average charge of the molecules at a given pH. This procedure takes into account the state of an ensemble of n identical solutions at a fixed pH, each containing a single molecule HA. At a given instant of time (t) the average charge can be calculated as follows:

$$\langle q \rangle_e = \frac{1}{n} \sum_{i=1}^{n} q_i(t) \quad \text{with} \quad n \gg 1, \tag{2.7}$$

where the subscript 'e' stands for ensemble average. Note the important difference between equations 2.6 and 2.7. The latter refers to the state of different molecules at the same moment in time while the former refers to the state of the same single molecule at different points in time. If $\langle q \rangle_{t_i} = \langle q \rangle_{\Delta t} = \langle q \rangle_e$, then it is said that the problem follows the ergodic principle, i.e. the time average of a single system (solution and molecule) is equal to the ensemble average [1]. Hereafter this approximation will be adopted and the notation $\langle q \rangle_{t_i} = \langle q \rangle_{\Delta t} = \langle q \rangle_e \equiv q$ will be used. Moreover, from the definition of acid ionization and dissociation constants the following relationship will be assumed:

$$q \equiv \langle q \rangle_{t_i} = \langle q \rangle_{\Delta t} = \langle q \rangle_e = -\alpha e = \left(\frac{-1}{1 + 10^{pK_a - pH}} \right) e, \tag{2.8}$$

where α is the degree of dissociation of the acid HA. The last term of this equation was deduced from equation 1.14. Equation 2.8 is plotted for the acid HA ($pK_a = 5$) in Figure 2.5 and for the conjugated acid BH^+ of base B ($pK_a = 9$) in Figure 2.6. They show the values of q at different pH values. Note that for HA q goes asymptotically to zero at low pH values and asymptotically to $-e$ at high pH values. Mathematically q will never be exactly zero or exactly $-e$; these are just the limiting values of q for the acid HA.

The average charge q, calculated in equations 2.5, 2.6, and 2.7, is related to the *effective mobility* of the analyte in free solution by the equation $\mu_{ef} = q/f = -\alpha e/f$, where f is the friction coefficient and α is the ionizing constant (fraction of ionized molecules). This is valid in concentration ranges where the ionization constant of

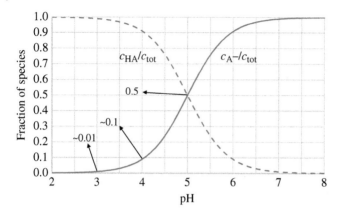

Figure 2.5 Molar fraction of the species of the acid HA ($pK_a = 5$) in relation to c_{tot} ($c_{tot} = c_{A^-} + c_{HA}$). The solid line represents the fraction of the charged species and the dashed line the fraction of the uncharged species.

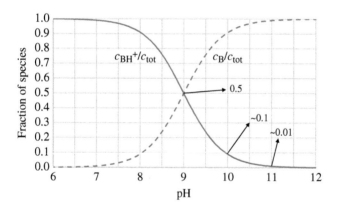

Figure 2.6 Molar fraction of the species of the conjugated acid BH^+ ($pK_a = 9$) of base B in relation to c_{tot} ($c_{tot} = c_B + c_{BH^+}$). The solid line represents the fraction of the charged species and the dashed line represents the fraction of the uncharged species.

HA is not affected by the concentration of HA. A similar calculation starting from Eq. 1.14 produces an expression for the effective charge of the conjugated acid of base B (monoprotonated base B). This is plotted in Fig. 2.6.

2.5 Ionic Limiting Mobility

Ionic limiting mobility (μ_o) is regarded as one of the fundamental parameters of an analyte (ion, charged molecule, or particle). It is expected to depend only on the solvent, temperature, and pH used. It is defined as $\mu_o = \lim_{I \to 0} \mu_e$ where I is the ionic strength of the solution. Remember that the ionic strength I is defined as half of the sum of the concentration of each ion times its valency squared: $I = (1/2)\sum_i c_i z_i^2$. Examples of entities that contribute to the ionic strength include the analyte itself and the mixtures of a weak acid and strong base (or a weak base and strong acid) that are commonly used to prepare the buffer solutions that fix pH. Moreover, these measurements are normally taken with the ion or small molecules at their full charge ($\mu_e = ze/f$) and at a reference temperature (usually 20 °C or 25 °C). However, there is a contradiction in this procedure: it is not possible to measure μ_e in an infinite dilution (zero ionic strength). Therefore, a few measurements are taken at decreasing values of I, that is, at decreasing buffer concentrations. The values of μ_e are then plotted against I and the limiting mobility value is determined by extrapolating the line of best fit to zero ionic strength, i.e. $\mu_o = \mu_e$ at $I = 0$. These measurements are easily taken for mono-atomic anions and cations, but organic molecules require more sophisticated approaches [6].

The dependence of the electrophoretic mobility (μ_e) on the ionic strength of the solution is given by the Debye, Hückel, and Onsager theory [7] or by the more elaborated theory of Onsager and Fuoss [8]. For the case of symmetric buffers with valency z, the limiting mobility (μ_o) can be calculated from the actual electrophoretic mobility (μ_e) and ionic strength (I) by this equation:

$$\mu_o = \mu_e + \frac{Az\sqrt{I}}{1 + 2.4\sqrt{I}}, \tag{2.9}$$

where A is a constant that must be estimated by using the line or curve of best fit for the set of points measured. This is made, for instance, using a buffer in the 5 mM to 100 mM range. The parameter 2.4 in equation 2.9 is related to the distance between the center of the ion under measurement and the start of the ionic atmosphere. It has been shown to be independent of the solute charge, temperature, and both counter-ion and co-ion used [9].

The hydron or hydronium ion (H_3O^+) has the highest electrophoretic mobility among all ions (cations and anions) and the hydroxyl ion (OH^-) has the second highest. This happens due to a jumping mechanism performed by the protons and directed by the applied electric field. Basically, one of the three protons that are bound to the oxygen atom of a hydronium ion (through sp^3 bonding orbitals), detaches and jumps (or tunnels - by passing through a potential barrier) to a nearby neutral water molecule. This lefts behind a neutral water molecule and creates a new hydronium ion. The overall effect is the same as if the hydronium ions itself were moving. A similar mechanism favors also the mobility of the hydroxyl ions, as they can be seen as a water molecule with the lack of a proton. This serves as a landing point for a proton from a neighbor water molecule. Again, the overall effect is the same as if the hydroxyl anions itself were moving, but in the opposite direction of the hydronium cations. It has been shown that this mechanism is dependent on the hydrogen-bonding rich environment of liquid water at room temperature [10].

2.6 Bands, Fronts, Peaks, and Zones

To highlight the different usages of the terms bands, fronts, peaks, and zones it must be remembered that analytical and preparative electrokinetic separations are usually conducted in only one spatial dimension, which is the direction of the applied electrostatic potential gradient. Therefore, the analytes' spatial concentration profiles are usually constant along the other two spatial coordinates. This means that they can be treated as a one-dimensional problem in most cases. The majority of the ESTs currently used are of the band type, in which a short sample plug is injected, separated, and then detected.

2.6.1 Bands and Peaks

The molecular distributions, which usually have a Gaussian profile, are called *bands*. Their concentration profiles can be observed along the separation paths, i.e. along the liquid core of capillaries, microchannels within microchips, or in lanes found within a slab. The bands are mathematically represented by the concentration c, which is given in terms of both a spatial coordinate (x) and time (t). The separations (runs) start at $t = 0$, when a high voltage is switched on. Therefore, the total concentration (c) of n analytes is the sum of the distributions $c_i(x, t)$, which is given as follows:

$$c(x, t) = \sum_{i=1}^{n} c_i(x, t). \tag{2.10}$$

Figure 2.7 illustrates the separation of a mixture of two components represented by c_1 and c_2, with respective migration velocities of v_1 and v_2. At $t = 0$ the components form part of the initial injected sample plug, with a narrow Gaussian (or normal) distribution along the x axis. This dynamics is governed by the diffusion equation (see Section 2.13 and Appendix E) with a drift term when the velocities and diffusion coefficients are constant in time and space. It is now possible to make a mathematical distinction between peaks and bands, which has been the source of some confusion in the literature. There are at least two ways to register the progress of the separation of the molecular distributions. One will show the bands and the other will produce an electropherogram with *peaks*. These methods are:

1. The snapshot method: Detection of the molecular concentration profile occurs at a fixed point in time (t_m) along the whole axis x, from $x = 0$ to L (the end of the capillary, microchip channel or slab). The detection results are mathematically represented by the intersection of the plane $t = t_m$ with the surface $c(x, t)$ of Figure 2.7, giving the spatial distribution of the bands at time t_m. Note that it is hard to access the bands in HPLC because of the stainless steel columns and/or the packing beads used, which makes detection of the spatial profiles harder.

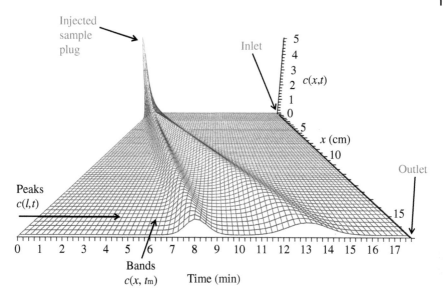

Figure 2.7 Visualization of function $c(x, t) = c_1(x, t) + c_2(x, t)$, which is the solution of the diffusion equation with drift terms or velocity terms. The injected sample plug has a Gaussian distribution centered at $x = 1$ cm along the x axis at the start ($t = 0$ min) of the run. As time progresses, from left to right, the successive spatial concentration profiles along the x axis (bands) can be seen from $x = 0$ (up) to $x \sim 15.2$ cm in the front. The separation of the two bands becomes apparent as the center of mass of one becomes more advanced in space than the other. The intersection of this surface ($c(x, t)$) with the plane $x = l$ defines the peaks of an electropherogram (arrow at $l = 15$ cm), while the intersection of this surface with the plane $t = t_m$ gives the band's spatial profile at migration time $t = t_m$ (arrow at $t_m = 6$ min). A snapshot taken at this time ($t_m = 6$ min) reveals the two bands while a snapshot taken at time $t = 10$ min shows only one band because the fastest band has already exited the system through the outlet.

2. The finish line method: A detector is placed at the fixed position l, the detection cell. The detector continuously monitors the passage of the bands, starting from $t = 0$ and continuing until the end of the run. The output is called an electropherogram (in electrodriven separations) or a chromatogram (in chromatographic separations). Electropherograms show a baseline and the peaks that result from the bands passing the line (or surface) at the fixed detection point (l). Therefore, both the electropherograms and chromatograms represent the intersection of the plane $x = l$ with the surface $c(x, t)$ of Figure 2.7.

Therefore, bands are the real objects given by $c(x, t_m)$ and peaks (in the electropherograms and chromatograms) are the products of detection signals that are produced when bands cross the finishing line, given by $c(l, t)$. The bands $c(x, t_m)$ present along the separation path x in the ESTs can be measured at any time t_m in many ways. One way is to take a photograph of the whole separation path (e.g.,

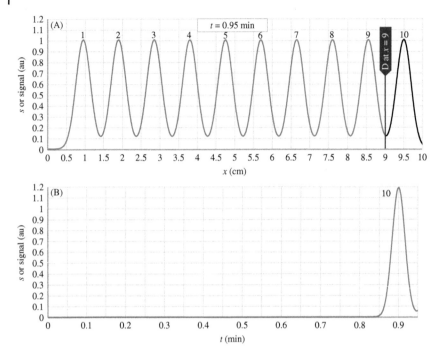

Figure 2.8 A separation simulation generated from a program written in a QuickBasic code. The total length of the separation path (capillary, microchip or slab) was defined as $L = 10$ cm, with a distance of 9 cm between injection and detection point ($l = 9$ cm). All bands have the same diffusion coefficient: $D = 0.015$ cm^2 min. (A) Snapshot of the bands at time $t = 0.95$ min. The numbers serve both as the identity of the bands and as an expression of their respective velocities in cm min^{-1}. Note that they are already separated at this time but only the front runner (band 10) has crossed the detection point. The run could be stopped here after taking a spatial scan of the bands. (B) Electropherogram of the same run showing the first peak (the fastest band), which has a velocity of 10 cm min^{-1}, and whose center of mass has just crossed the detector (D) – detection point positioned at $x = 9$ cm.

the slab made of a gel or a capillary in isoelectric focusing). A second way is to use a confocal LIF detector driven by a servo-motor that quickly scans the separation path (open or toroidal) at time t_m and saves the fluorescence signals with their respective spatial addresses provided by the servo-motor. This gives the spatial band profile at time t_m. In HPLC the baseline with peaks is the only data available because the bands cannot be "seen" through the stainless steel columns. In the ESTs both the electropherogram and the bands' spatial profiles along the separation media are generally accessible.

Figures 2.8, 2.9, and 2.10 shows a simulation of the separation of ten analytes that was obtained using a programming language (QuickBasic) to calculate the

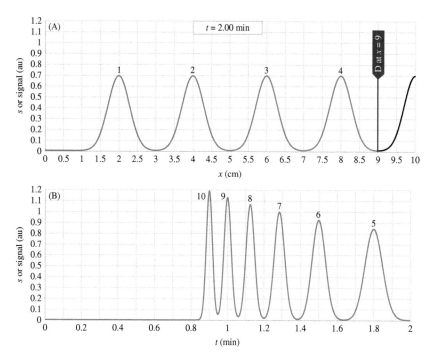

Figure 2.9 The same hypothetical run as in Figure 2.8, but this time it is shown after a run time of 2 min. (A) Snapshot of the bands at time $t = 2$ min. The six fastest bands have already crossed the detection point and exited the outlet of the separation platform (capillary, microchip, or slab). The four remaining bands are already excessively resolved at this time. (B) Electropherogram after run time $t = 2$ min, showing the six fastest analytes with velocities of 10, 9, 8, 7, 6, and 5 cm min^{-1}.

time evolution of the center of mass of the bands, standard deviations, and signal at the detector position. All ten constituents have the same diffusion coefficient ($D = 0.015$ cm^2 min) and their velocities are 1, 2, 3, 4, 5, 6, 7, 8, 9, and 10 cm min^{-1} (the same number is used to label them). The detector (which produces the electropherogram) is placed at $x = 9$ cm, and the total length of the separation track (capillary, microchip, or slab) is 10 cm. The sample is injected at $x = 0$ with a narrow width (negligible compared to the final standard deviation) and the high voltage is immediately turned on at $t = 0$ to start the separation process. Figure 2.8A shows the spatial profile of the bands at time $t = 0.95$ min and Figure 2.8B shows the respective electropherogram at this point in time. Note that at time $t = 0.95$ min all of the bands are already separated from each other (*resolution* is defined in Section 3.5.2.1). However only one peak (the front runner) appears in the electropherogram at this time as much more time is required for all the other peaks to be recorded in the electropherogram.

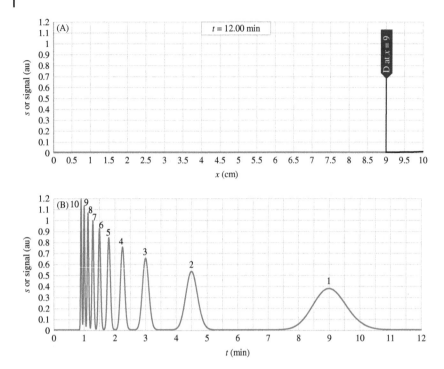

Figure 2.10 End of the hypothetical run of Figure 2.8. (A) The separation path is now empty because all bands have already exited at $x = 10$ cm. (B) Electropherogram after time $t = 12$ min, showing all peaks. Note that the velocity difference between any two neighboring bands is exactly 1 cm min^{-1}, nevertheless the time differences between the last peaks on the electropherogram are enormous. Moreover, all bands present the same area during the run, however the peaks in the electropherogram exhibit large differences in area caused by the different speeds at which they cross the detection point. Finally, the peaks are asymmetrical in the electropherogram only because they spread, due to molecular diffusion, while crossing the detection point. This asymmetry is more visible on peaks 2 and 1 as, for example, their right side tails are higher than their left side tails at equal distances from the positions of the maxims.

Figure 2.9A shows an intermediate situation: the six faster bands have already crossed the detection point and the four slower bands are still on their way to the detection point. The electropherogram shows that the slower peaks are wider (Figure 2.9B). This is because of the combination of two distinct factors: the bands that arrive later have more time to diffuse along the separation track, and the slower bands take more time to cross the detection point.

Figure 2.10A shows an empty track (capillary, microchip, or slab) at the end of the run, when all the bands have already exited the separation track. At this point, the electropherogram (Figure 2.10B) has a record of all constituents (ten peaks). Note that the peaks in the electropherogram are always a little asymmetric, visible by comparing the left and right tails of peak 1. This is because of at least one unavoidable factor: band diffusion during their passage through the detection point.

Other important parameters, besides the velocity v_i of each analyte, are the standard deviations (σ_i) and variances (σ_i^2) of the bands and peaks. The standard deviations of the bands have units of length and the peaks have units of time (minutes or seconds). The conversion $\sigma_i(x, t_m) = \sigma_i(l, t) \times v_i$ is commonly found within the literature. The error associated with this transformation is small if the transit time is small, $\sigma_i(l, t) \ll t_m$. The time dependence of standard deviations and variances is discussed in Section 2.13.

2.6.2 Fronts

In the *frontal techniques* the samples are continuously fed into the separation medium during the runs, or at least for a significant fraction of the total run time. Therefore, one reservoir is filled with the sample while the separation track and the opposite reservoir are filled with a buffer solution. For high quality separations, preference should be given to buffer solutions with pH and conductivity values that are similar to those of the sample. Additionally, the cations and anions of the buffer should have similar mobilities to those of the sample; however, their detection characteristics should be distinctly different. After running for a while under the influence of the applied electric field, the analyte concentration profiles resemble long plugs with a boundary at the "frontal" edge. The resulting signal, detected at the crossing line, is shaped as a staircase with varying distances between each step (nature of the analyte) and differing step heights (concentrations). However, these steps do not have sharp edges, they are rounded due to molecular diffusion and other dispersion mechanisms.

The time derivatives of these staircase-like electropherograms give peaks with respective migration times (inflection points of the frontal boundaries), standard deviations, heights, and widths. Conversely, the integration of the signal of electropherograms produces the signal that would be detected if a frontal technique is used to separate the mixture. The same happens with the bands: the integration over space of the signal produced by a scanner used to detect the bands will produce the staircase signal that would be produced by the scanner if the frontal technique is used to separate that particular mixture. Also, the converse is also true, the spatial derivative of the signal produced by a scanner of the separation of a frontal technique will produce the bands that would be observed by the scanner if a band technique is used to separate them. Most of the performance parameters discussed in Sections 3.5 and 4.9 and given in the Tables of Chapter 5 apply to these band-like and peak-like signals that are obtained when taking the derivatives of the signals produced by the frontal techniques. Moreover, it is very easy to make the interconversions frontal to bands, bands to frontal, frontal to peaks, and peaks to frontal using a spreadsheet such as Microsoft Excel™. These frontal techniques, while very suitable for fraction collection, are not commonly used for some reason.

Free flow electrophoresis [11] (FFE) is not a frontal technique, despite the continuous hydrodynamic feeding it utilizes. It is a band technique because the electrokinetic separation occurs orthogonal to the feed direction. This technique is even better for fraction collections and preparative separations, despite the fact that it has lower resolutions than band and frontal techniques.

2.6.3 Zones

Isotachophoresis [12–16] (ITP) is a *zone* technique, but it has a peculiar separation mechanism. In this technique the zones are usually longer than the bands in the other ESTs, with exception of the frontal techniques. Here, a *leading electrolyte* is used to fill the separation medium and the destination reservoir, while a *terminating electrolyte* is added to the inlet reservoir. When anions are analyzed, then the anion of the leading electrolyte must have a higher mobility than the fastest anion of the sample and the cation of the terminating electrolyte must have a lower mobility than the slowest cation of the sample. This leads to strong electric field distortions and a *sui generis* separation mechanism. When the system reaches equilibrium the analytes move as juxtaposed plateaus, a train of zones with the same velocity, aligned in order of decreasing mobilities. In this case zone width is related to the concentration of the respective analyte and the order of arrival is related to the identities of the analytes. This may be intriguing for anyone familiar with chromatography, therefore the mechanism is explained in more detail in Section 2.15.6.

The denominations of bands and zones were used interchangeably at the beginning of the development of capillary electrophoresis. However, the term zone is most suitable if the spatial distributions of the analytes have rectangular profiles and are long compared to the ID of capillaries and microchannels. If the spatial profiles of the analytes are Gaussian (according to the *normal distribution*) then the denomination of band is more appropriate.

2.7 The Isoelectric Point

The isoelectric point (IEP) is another fundamental property of a molecule, just like its molecular mass, limiting mobility (discussed in Section 2.5), pK_a values, absorption spectra, and structural formula (the way the atoms in a molecule are arranged in space).

2.7.1 Isoelectric Point of Molecules

A given molecule has an IEP denoted by pH(I) if, and only if, it possesses the following two properties:

1. At pH = pH(I) the molecule has a null electrophoretic mobility.
2. If the pH is raised by a small amount (δpH) to pH = pH(I) + δpH, then the molecule assumes a small average charge of δq. If the pH is lowered by a small amount (δpH) to pH = pH(I) − δpH, then the molecule assumes a small average charge of −δq. The same statement can be made for small changes of mobility, by substituting δq with $\delta\mu$.

For the above conditions to be fulfilled by a given molecule, the molecule must contain at least one acidic group and one basic group. The molecules discussed in Section (2.4), which only have one acid (Figure 2.5) or one basic group (Figure 2.6), do not have an IEP because their mean charge (q) will never cross the $q = 0$ axis at any pH.

The property 1, listed above, is very interesting because at pH = pH(I) many species (see Figure 1.4) may co-exist. This can occur with molecules possessing more than one acidic or basic group, or when the pK_a of the basic group of a molecule is very close to the pK_a of its acidic group. It is useful to examine two distinct extreme situations: (i) when the pK_a of the acidic group is lower than the pK_a of the conjugated acid of the basic group, and (ii) when the pK_a of the acidic group is higher than the pK_a of the conjugated acid of the basic group. Both situations have an IEP, but in the first case the two groups are predominantly charged at pH = pH(I) (see Figure 2.11) and in the second case the two groups are predominantly uncharged at pH = pH(I) (see Figure 2.12).

The first case is observed in a class of molecules called zwitterions. Most of the buffers known as *De Good's buffers* fall within this category, and they are good buffers for ESTs. Polymers containing repeated ionic groups, each possessing both a positive and a negative charge, as well as pK_a values that are separated from each other by approximately one unit or less, are also very interesting buffer choices in ESTs as they exhibit a high buffer capacity at pH values that are close to their pH(I). Therefore, they contribute minimally to the solution's conductivity. When these two charged groups are incorporated into the pendant group of ampholytic polymers then they are called zwitterionic polymers.

The second case (as shown in Figure 2.12), when both groups are predominantly uncharged at pH = pH(I), is observed when the same molecule contains a very weak acidic group and a very weak basic group. If the pK_a of the conjugated acid of the basic group is at least two units smaller than the pH(I), and the pK_a of the acidic group is at least two units larger than the pH(I), then at pH = pH(I) the predominant species will be the one that has uncharged acid and basic groups. In this case each basic and acidic group of the molecule will have a charge of zero, giving the molecule an overall net charge of zero and consequently a null electric mobility when in free solution electrophoresis. Note that a molecule of this type does not assume a zwitterionic form at the IEP, nevertheless it fulfills the two

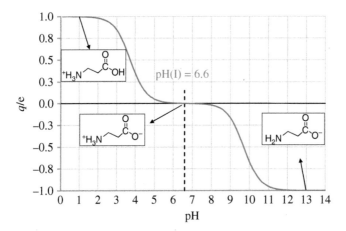

Figure 2.11 Average charge (ensemble average or time average) of β-alanine, which is the same structure studied in Figure 1.8. It has a carboxylic group ($pK_a = 3.63$), an amine group ($pK_a = 9.6$), and an isoelectric point of 6.6 (pH(I) = 6.6). The pK_a of the acid group is lower than the pK_a of the conjugated acid of the basic group. Therefore both functional groups are predominantly charged at pH values around pH(I), but the overall net charge is zero at pH = pH(I). The predominant species at pH = 1, 6.6, and 13 are shown in the boxes.

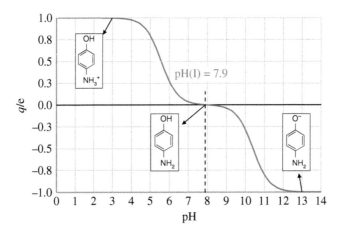

Figure 2.12 Average charge (ensemble average or time average) of 4-aminophenol, which has a phenol group ($pK_a = 10.30$), an aromatic amine ($pK_a = 5.48$) and an isoelectric point of 7.9 (pH(I) = 7.9). Note that the pK_a of the acidic group (phenol) is higher than the pK_a of the conjugated acid of the basic group (aromatic amine). In this case the molecules are predominantly uncharged at pH values of around pH(I). The predominant ionic forms at pH = 3, 7.9, and 13 are shown in the boxes.

necessary properties to have a IEP. Despite the low conductivity of these molecules at values of pH that are close to their pH(I) values, these molecules usually are less soluble in aqueous solutions than zwitterionic molecules.

Dozens of ESTs have been used for the experimental measurement of the IEP of organic molecules and especially proteins, [17] including compact microchip based free-flow electrophoresis [18] and with the pH of the electrolyte solution modulated in time rather than in space for faster measurements.[19] There are also dozens of software and web utilities dedicated to predict the pH(I) of organic molecules and proteins. They are fed with the pK_a values of the molecules. Fortunately, there are at least fourteen methods for determining the acid dissociation constants of organic molecules [20].

2.7.2 Isoelectric Point of Nano and Microparticles

In the case of colloids consisting of aggregates of macromolecules, cells, and nano- and microparticles that are dispersed in aqueous solutions, the following two situations are encountered. Either the suspended particles do not have ionizable groups on their surfaces or the suspended particles do have acidic and/or basic groups located on their surfaces. Examples of the first case include microdroplets of oils made from alkanes or microparticles of polyolefins dispersed in aqueous solutions. Even in this case the suspended microdroplets or particles may acquire an interface potential when placed into any aqueous solution, even pure water. This happens due to the asymmetric distribution of the OH^-, H_3O^+, and other ions at the interface of the water and the microparticles or microdroplets. This also happens, for instance, with microbubbles of air (or any inert gas) dispersed in water. Note that the electrophoretic mobility of the microparticles or microdroplets will be zero if the electrostatic potential at the shear plane is zero or if the viscosity at this shear plane is very high (see Section 2.9 and Appendix D). The second case occurs when a colloid has acidic and/or basic groups at their surfaces. Quartz microparticles, for instance, have silanol groups that ionize to SiO^- at pH > 5. Similar to the first case, the electrophoretic mobility will only be zero in aqueous solutions if the ζ potential is zero or if the viscosity at the slipping plane is very high. Finally, for the general case of all particles, it is correct to say that they will always have an IEP if the following two conditions are fulfilled:

1. At pH = pH(I) the particle exhibits null electrophoretic mobility ($\mu_e = 0$).
2. If the pH is raised by a small amount (δpH) to pH = pH(I) + δpH then the particle assumes a small zeta potential ($\delta\zeta$), and if the pH is lowered by a small amount (δpH) to pH = pH(I) − δpH then the particle assumes a small zeta potential ($-\delta\zeta$) at the shear plane, which will always have an opposite sign to the zeta potential of the first case. The same statement can be made for small changes of the electrophoretic mobility, by substituting $\delta\zeta$ for $\delta\mu_e$.

Attempts to measure the zeta potential through the measurement of particle mobility may not be feasible if the particles contain polymers that are adsorbed on their surfaces, as this will disturb or even completely suppress their electrical mobility.

2.8 Turbulent and Laminar Flow

Turbulent and laminar flows have important practical implications in the separation sciences. Here only *Newtonian* liquids that are homogeneous and incompressible will be considered. Turbulence is characterized by the convection of mass, which is a very efficient band dispersing mechanism. Therefore, laminar flows are generally preferred over turbulent flows in separation sciences and they will be examined in more details.

2.8.1 The Driving Forces of Fluid Flow

There are at least five different ways to promote the movement of fluids along microtubes, microchannels, and conduits: electroosmosis, centrifugal forces, gravity, pressure, or using capillary action. Electroosmosis, which is driven by the applied electric field, is discussed in Section 2.9 and has an odd (piston-like) velocity profile. On the other hand centrifugal, gravity, pressure, and capillary action driven flows exhibit their highest velocities at the central line of the conduit. These velocities then fall, with a parabolic profile, to zero at the wall.

Gravimetric or gravity driven flows depend on differences in liquid levels between reservoirs (inlet and outlet), local gravity acceleration, conduit geometry, density, and the dynamic viscosity of the fluid. These flows usually occur when ESTs are mistakenly performed with unleveled reservoirs. However they are also intentionally used to prevent band leaking into the reservoirs in the toroidal layout (see Chapter 4).

Centrifugal force is an apparent outward force that can be observed in a rotating reference frame. It depends on the frequency of rotation, distance from the rotation axis, conduit geometry, and both the density and dynamic viscosity of the fluid. In this case the observed liquid flow always occurs in the outwards direction.

Pressure driven velocity profiles depend on the applied pressure gradient, conduit geometry, and liquid dynamic viscosity. Pressure gradients are used in the ESTs for many purposes, for instance: to inject the sample into the separation medium, to drive the bands to the detection point in the iosoelectric focalization separation mode, in pressure assisted capillary electrophoresis [21], in pressure-assisted capillary electrophoresis frontal analysis [22], and to mitigate thermal band broadening when large temperature gradients are observed across the separation medium [23]. (see the use of Poiseuille counter-flow to mitigate thermal peak broadening in Appendix F).

The capillary action is a phenomenon that fills microchannels and capillaries with liquids when they are laid horizontally. It acts also against gravity until a certain limit of ΔH, the liquid level difference. It depends on microchannel radius, liquid–air surface tension, and the contact angle of the liquid and the material of the microchannel or capillary (a measure of intensity of cohesive forces between the liquid and the material).

2.8.2 Turbulence

Fortunately, there is an efficient criterion to predict whether a given fluid flow will be turbulent or not. This is given by the number R, according to equation 2.11, which is called the Reynolds number (in recognition of the British scientist and mathematician Osborne Reynolds, 1842–1912). When R is smaller than ten ($R <$ 10) the probably of finding a segment of length d with a turbulent flow, even in the transient regime, is negligible. To put some real numbers into equation 2.11, consider a microtube (capillary), with d = ID = 50 μm, which is filled with pure water at a velocity of 10^{-3} m s^{-1}. In this case $R = 0.05$, which is much smaller than ten. This property is and advantage of the ESTs conducted in capillaries, microchannels, and slab of gels, as turbulence is common in other separation techniques such as chromatography, centrifugation, distillation and most of the field flow fractionation techniques.

The Reynolds number (R) is given by:

$$R = \frac{\rho v d}{\eta}, \tag{2.11}$$

where ρ is the density of the fluid, v its velocity, d the diameter of the cylinder, and η the dynamic viscosity.

In conclusion, the pressure driven velocity profiles of liquids inside capillaries and microchannels exhibit a laminar flow in almost all practical situations of the ESTs. Therefore, it is important to have a detailed picture of the velocity profiles of the liquids (usually aqueous solutions) across these micro conduits, i.e. the separation media. The velocity profiles at the entrance and exit will be omitted in the next sections and only the steady state regime will be considered.

2.8.3 Laminar Flow in Cylindrical Capillaries

2.8.3.1 Pressure Driven Flow

For the capillary and coordinate system shown in Figure 2.2 the pressure driven velocity in the x direction only depends on r. In this case the Navier–Stokes equations, for an incompressible fluid in the steady state, with a constant and uniform viscosity, reduces to the simple form shown by equation 2.12.

The Navier–Stokes equation for an unidimensional laminar flow of an incompressible fluid with constant viscosity in a cylindrical pipe:

$$\frac{1}{r}\frac{\partial}{\partial r}\left(r\frac{\partial v}{\partial r}\right) = -\frac{1}{\eta}\frac{\partial P}{\partial x}, \tag{2.12}$$

where P is the pressure and η is the dynamic viscosity.

The solution of the steady, laminar, axisymmetric flow in a circular pipe of radius a, length L, with $v(a) = 0$, is given by:

$$v(r) = \frac{1}{4\eta}\frac{\Delta P}{L}a^2\left(1 - \frac{r^2}{a^2}\right), \tag{2.13}$$

where ΔP is the pressure difference and $L \gg a$. Note that the term in parenthesis is a dimensionless parameter that changes from zero to one. This result is shown in Figure 2.13 as a top view in Cartesian coordinates. This parabolic velocity profile is known as Poiseuille flow (in recognition of the French medical doctor and

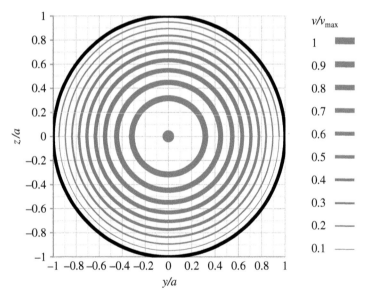

Figure 2.13 Top view of the Poiseuille velocity profiles shown as level contours in Cartesian coordinates, with $-a < y < a$ and $-a < z < a$, where a is the inner radius of the capillary. The pressure driven velocity occurs in the x direction, has its maximum (v_{max}) at the center line and decreases in the surrounding surfaces (lamins). The velocity is proportional to the thickness of the lines and the level contours are shown for $v/v_{max} = 1.0, 0.9, 0.8, 0.7, 0.6, 0.5, 0.4, 0.3, 0.2,$ and 0.1.

physicist Jean-Leonard-Marie Poiseuille, 1797–1869. The cgs unit of dynamic viscosity, Poise, was also named after him).

The maximum velocity (v_{max}), which occurs along the center line ($y = 0$ and $z = 0$), and the total volumetric flow rate ($\Delta Vol/\Delta t$) are given by:

$$v_{max} = \frac{1}{4\eta}\frac{\Delta P}{L}a^2 \quad \text{and} \quad \frac{\Delta Vol}{\Delta t} = \frac{\pi}{8\eta}\frac{\Delta P}{L}a^4. \tag{2.14}$$

2.8.3.2 Gravity Driven Flow

The maximum velocity and volume of liquid that flows through a cylindrical microtube during the time interval Δt as a result of the level differences can be calculated from equation 2.14. The term ΔP must be replaced by the pressure exerted by the column of liquid of the levels difference (ΔH):

$$v_{max} = \frac{1}{4\eta}\frac{\rho\,g\,\Delta H}{L}a^2 \quad \text{and} \quad \frac{\Delta Vol}{\Delta t} = \frac{\pi}{8\eta}\frac{\rho\,g\,\Delta H}{L}a^4, \tag{2.15}$$

where ρ is the density of the fluid, g is the local gravity acceleration, and ΔH is the liquid levels height difference if the microtube ends are immersed into two reservoirs or connected as communicating vessels.

2.8.3.3 Capillary Action

The capillary action may fill completely tiny capillaries with liquids when they are laid horizontally. It acts also against gravity when these tiny capillaries stand vertically, but only until a certain limit of ΔH. This maximum pressure that capillary action can sustain, inside a cylindrical capillary with inner radius of a, is given by:

$$\Delta P = \rho g \Delta H = \frac{2\gamma\cos\theta}{a}, \tag{2.16}$$

where ρ is the liquid density, g is local gravity acceleration, ΔH is the liquid level difference, γ is the liquid–air surface tension (Newtons/meter), and θ is the measurable contact angle of the liquid and the material of which the microchannel or capillary is made (a measure of the intensity of cohesive forces between the liquid and the material).

When a capillary is laid in the horizontal position, such that $\Delta H = 0$, then the maximum velocity (v_{max}) and rate at which the liquid flows ($dVol/dt$) into a cylindrical microtube can be calculated using equations 2.16 and 2.14. The result, of pure capillary action, is given by:

$$v_{max} = \frac{1}{2\eta}\frac{\gamma\cos\theta}{L'}a \quad \text{and} \quad \frac{dVol}{dt} = \frac{\pi}{4\eta}\frac{\gamma\cos\theta}{L'}a^3, \tag{2.17}$$

where L' is the length of the capillary that was already filled with liquid (and not the total length of the capillary). The remaining length of the capillary ($L - L'$) is supposed to be filled with air. Equation 2.17 is not valid for $L' = 0$ and $L' \simeq 0$ because the role of the liquid's mass inertia is not taken into account in equation 2.17 for these values of L'. Note that v_{max} and $dVol/dt$ are high at the beginning when $L' \simeq a$ and become very small when $L' \simeq 10^3 a$ or longer.

2.8.4 Laminar Flow in Microchannels

2.8.4.1 Pressure Driven Flow

The square and rectangular microchannels are currently the most popular geometries used in the microchip platforms (microchip electrophoresis). They are produced using soft-lithography, hot-embossing, and 3D printing, among other techniques. There are also flexible fused-silica microtubes (capillaries) that have a square shape, i.e. they have an inner lumen with width and height of 50, 75, or 100 μm. Externally the fused quartz is also square with a width and height of ~365 μm (the polyimide gives a rounded shape). They are not well known and are not often used, but enable improvements of the detection systems and the cooling setups.

The Cartesian coordinate system shown in Figure 2.1 must be used to calculate the pressure driven velocity profiles in these square conduits. As in Section 2.8.3, the pressure driven flow is considered in the steady state, with only one velocity component in the x direction, for an incompressible fluid, and with constant and uniform viscosity. In this case the Navier–Stokes equations reduce to equation 2.18.

$$\frac{\partial^2 v}{\partial y^2} + \frac{\partial^2 v}{\partial z^2} = -\frac{1}{\eta}\frac{\partial P}{\partial x}.$$ (2.18)

This second order linear non-homogeneous differential equation is known as Poisson's equation (in recognition of the French mathematician and physicist Siméon Denis Poisson, 1781–1840) with a non-homogeneous term (the right hand term). Velocity profiles can still be calculated in these cases but they do not have a simple analytical solution, apart from expansion in eigenfunctions with an infinite sum. Due to symmetry reasons the Cartesian coordinate system shown in Figure 2.1 is the most advantageous in this case, as a sum of cosine functions can be used to solve the problem in both y and z directions. This gives the simplest solution. In this case the solution of equation 2.18, with the boundary conditions $v(w, z) = v(-w, z) = v(y, h) = v(y, -h) = 0$, is given by:

$$v(y, z) = \frac{1}{\eta}\frac{\Delta P}{L}h^2 \sum_{i=1}^{\infty} \frac{(-1)^{i+1}}{\alpha_i^3} \cos(\alpha_i z/h)\left[1 - \frac{\cosh(\alpha_i y/h)}{\cosh(\alpha_i \alpha)}\right],$$ (2.19)

where $\alpha_i = (2i - 1)\pi/2$ with $i = 1, 2, \dots$. The height of the channel is given by $2h$, the width by $2w$, and the aspect ratio ($\alpha = w/h$) is assumed to be equal to or larger than one ($\alpha \geq 1$). This sum converges very quickly, considering only the first ten terms it gives an error of less than 0.01% if $\alpha \leq 10$. Figures 2.14, 2.15, and 2.16

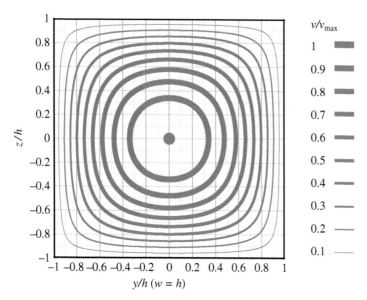

Figure 2.14 Top view of the laminar fluid velocity profiles shown as level contours in Cartesian coordinates across a square conduit (square microchannel or square fused silica capillary). The pressure driven velocity occurs in the x direction (coming out of this page), has its maximum ($v_{max} = 0.29\Delta P\, h^2/(\eta L)$) at the center line and decreases radially, as shown by the surrounding "isovelocity surfaces" (lamins). The velocity is proportional to the thickness of the lines and the level contours are shown for $v/v_{max} = 1, 0.9, 0.8, 0.7, 0.6, 0.5, 0.4, 0.3, 0.2$, and 0.1.

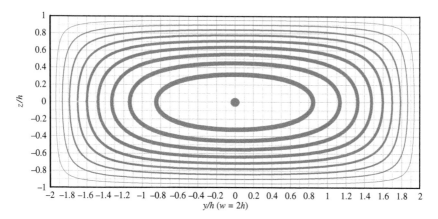

Figure 2.15 Top view of the fluid velocity profiles shown as level contours in Cartesian coordinates across a rectangular conduit (rectangular microchannel or rectangular fused silica capillary) with an aspect ratio of two ($\alpha = w/h = 2$), i.e. the width is two times the height. The legend is the same as in Figure 2.14.

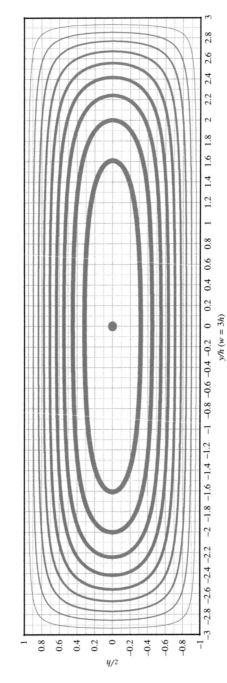

Figure 2.16 Fluid velocity profiles shown as level contours in Cartesian coordinates across a rectangular conduit with an aspect ratio of three ($\alpha = w/h = 3$). The legend is the same as shown in Figure 2.14.

Table 2.1 Values of the dimensionless parameters γ and β of equation 2.21, given as a function of the aspect ratio ($\alpha = w/h$) of the square ($\alpha = 1$) and rectangular ($\alpha > 1$) microchannels, capillaries or conduits in general.

$\alpha = w/h$	γ	β
1	0.294685	0.473233
2	0.455487	1.506714
3	0.490730	2.589836
4	0.498073	3.675199
5	0.499599	4.760661
6	0.499917	5.846125
7	0.499983	6.931591
8	0.499997	8.017056
9	0.499999	9.102523
10	0.500000	10.18799
> 10	$\simeq 1/2$	$1.085\,\alpha - 0.65$

show, respectively, the results of the velocity profiles as level contours for a square ($w = h$), a rectangle with $w = 2h$ and a rectangle with $w = 3h$. These are the typical values used in microchip electrophoresis and will be examined in detail.

The maximum velocities (v_{max}) and flow rates ($\Delta\text{Vol}/\Delta t$) in these cases are given by:

$$v_{max} = \frac{\gamma}{\eta}\frac{\Delta P}{L}h^2 \quad \text{and} \quad \frac{\Delta\text{Vol}}{\Delta t} = \frac{\beta}{\eta}\frac{\Delta P}{L}h^4. \tag{2.20}$$

The dimensionless parameters γ and β depend on the aspect ratio $\alpha = w/h$ and are tabulated in Table 2.1. Note that v_{max} reaches a maximum value around $w/h = 10$. Therefore, increasing the ratio w/h beyond $w/h = 10$, for a fixed value of h, does not increase the velocity along the center line of the rectangular conduit. However, the flow rate (Q) increases linearly with w/h when $w/h > 10$ and h is a fixed value.

Comparing equations 2.14 and 2.21, as well as Table 2.1, is possible to note that $1/4 < 0.294685$ and $\pi/8 = 0.392699 < 0.473233$. This is because a cylinder with diameter $2h$ fits inside a square tube with a width of $2h$, with some space left at the four corners. This makes both the fluid velocity at the center line and the total flow rate to be larger in the square tube than in the cylindrical tube.

2.8.4.2 Gravity Driven Flow
As in the previous subsection, ΔP in equation 2.21 must be replaced by the pressure exerted by the column of liquid of the levels height difference. This gives the

maximum velocity and the volumetric flow rate driven by gravity only:

$$v_{max} = \frac{\gamma}{\eta} \frac{\rho g \, \Delta H}{L} h^2 \quad \text{and} \quad \frac{\Delta \text{Vol}}{\Delta t} = \frac{\beta}{\eta} \frac{\rho g \, \Delta H}{L} h^4. \tag{2.21}$$

The volume that flows in the time interval Δt is proportional do the density of the liquid, gravity, and level difference, and is inversely proportional do the dynamic viscosity and conduit length. The parameters γ and β are given in Table 2.1 and depend on the aspect ratio (w/h).

2.9 Electroosmosis

2.9.1 EOF in Cylindrical Capillaries

The contact of aqueous solutions with the inner surface of a capillary, the walls of a microchannel or the bottom and lateral sides of an electrophoresis chamber or tank (for slab electrophoresis), causes an orthogonal displacement of charges in relation to these surfaces. The result is a thin sheet with a net surface density of charge on the solid dielectric surface and a thicker diffused layer within the aqueous phase with a net charge of the opposite sign. The thickness of this diffuse layer is the result of the interplay of two factors: the thermal energy that pulls the ions away from the interface and the electrostatic attraction that tries to bring the ions back to the interface. This interplay is governed by the *Poisson–Boltzmann equation* (see Appendix D). Moreover, due to the permanent dipole–charge inter-actions, layers of immobilized water molecules containing some of the diffused charges are held static, forming an immobile blanket over the solid dielectric surface (see Figure 2.17). This is the so-called *Stern layer* and it is important because it goes from the dielectric interface, where the resulting electrostatic potential is at its maximum [$\Psi(a) = \Psi_o$], to the shear plane (interface between the Stern layer and the fluidic phase), where the electrostatic potential is $\Psi(a') = \zeta$ (electrokinetic potential or "zeta" potential). In the fluidic phase the electric potential goes from $\Psi(a') = \zeta$ to $\Psi(0) = 0$. The Stern layer ($a' < r < a$) is only a few dozen nanometers thick, while the thickness of the diffuse layer, which is given by $1/\kappa$ (the distance over which the potential drops by a factor of e^{-1}) is a few hundred nanometers thick and depends on ionic strength, relative permittivity, temperature, and other factors (see Appendix D for details).

Application of the electric field E along the above mentioned surfaces (along x) is necessary to promote migration of the analytes. This also pulls the charges of the diffuse layer. If $1/\kappa$ is much smaller than the capillary ID or the microchannel width then the whole liquid phase will be displaced at a constant velocity – like merchandise on a conveyor belt. This causes a smooth displacement of the entire

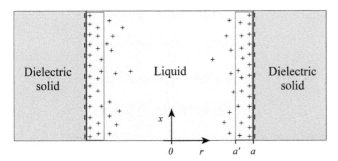

Figure 2.17 Illustration of the electrical double layer at the capillary wall. The net charge of the diffuse layer is responsible for the EOF. The ions and molecules in the red rectangles ($a' < r < a$) are immobile (called the Stern layer). However, the net charge (represented by the cations) in the fluidic phase, $0 < r < a'$, move all in the same direction (to the cathode) when the high voltage is turned on. This in turn moves the whole liquid from the anode to the cathode due to the drag effect. For this illustration $\kappa a \sim 5$. κa is usually much higher than one thousand for the microchannels and capillaries normally used, leading to an odd velocity profile. Figure 2.18A shows the unique piston-like velocity profile of electroosmosis.

liquid phase, with a constant velocity profile across the whole liquid, except for very close to the fluidic phase–Stern layer interface where the velocity falls to zero. The dependence of velocity on r is calculated in Appendix D and is given by:

$$v(r) = -\frac{\zeta E}{\epsilon_r \epsilon_o \eta} \left[1 - \frac{I_o(\kappa r)}{I_o(\kappa a)} \right] , \qquad (2.22)$$

where ζ is the electrostatic potential at the shear plane, i.e. at $r = a'$ (see Figure 2.17). For quartz and glasses this may reach -50 mV in alkaline pH and 25 °C. Note that the flow points from the cathode to the anode in this case. I_o is the *Bessel function of first kind and null order* and $I_o(0) = 0$. Unlike the parabolic flow profile observed in pressure driven flows and studied in Section 2.8, this fluid flow has a piston-like velocity profile if $\kappa a > 100$. This is shown in Figure 2.18 and the velocity along the center line is given by:

$$v_{eo} = \mu_{eo} E = -\frac{\zeta E}{\epsilon_r \epsilon_o \eta} = -\frac{\zeta V}{\epsilon_r \epsilon_o \eta L} , \qquad (2.23)$$

were μ_{eo} is the electroosmotic mobility, ϵ_r is the relative permittivity of the liquid, ϵ_o is the vacuum permittivity, and η is the dynamic viscosity of the liquid. The derivation of equation 2.23 is given in Appendix D. Figures 2.13–2.16 and 2.18-B show the pressure driven velocity profiles with small Reynolds' numbers (laminar flow), which is usually the case in capillaries and microchannels, and Figure 2.18A shows the velocity profiles of electroosmosis.

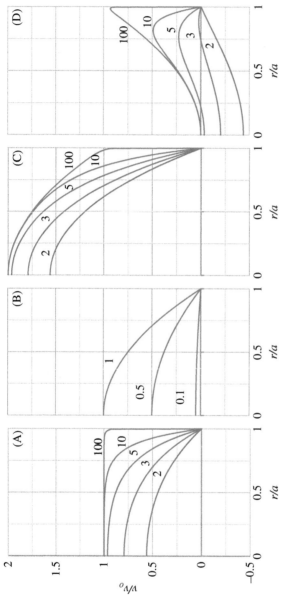

Figure 2.18 EOF and Poiseuille velocity profiles for cylindrical microtubes (capillaries) from the center ($r = 0$) to the wall ($r = a$). Graph A shows the velocity profile of the liquid phase when only electroosmosis is present (the Poiseuille term is null). The numbers represent the κa values and $v_o = -\zeta E/(\epsilon_r \epsilon_o \eta)$. In graph B only the pressure driven flow exists and the EOF is null. Here, $v_o = \Delta P \, a^2/(4\eta L)$ with $\Delta P = 1$, 0.5, and 0.1 given in units of pressure represented by $\eta L/a^2$. Graph C shows the velocity profiles when both the EOF and Poiseuille profiles are present and acting in the same direction. The parabolic velocity profile $\Delta P = 1$ of graph B is added to the EOF profiles of graph A in this case. Graph D occurs when both the EOF and Poiseuille term are present but acting against each other, i.e. the EOF profiles of graph A are subjected to the Poiseuille velocity term $\Delta P = 1$ of graph B, acting in opposite directions to each others.

2.9.1.1 Volumetric Flow Rate

The maximum velocity of the fluid and the volumetric flow rate are easily calculated from equation 2.23 if $\kappa a \gg 100$, which is usually the case.

$$v_{max} = \frac{\mu_{eo}V}{L} \quad \text{and} \quad \frac{\Delta \text{Vol}}{\Delta t} = \frac{\pi \mu_{eo}V}{L}a^2, \tag{2.24}$$

where a is the radius of the cylindrical microtube and V is the electrostatic potential difference applied between the ends of the microcylinder with length L.

2.9.1.2 Combined Pressure Driven Flow and EOF

The velocity profile of the liquid inside a capillary that results from the combined effect of pressure driven and electroosmosis is given by combining equations 2.23 and 2.13:

$$v(r) = \frac{1}{4\eta}\frac{\Delta P}{L}a^2\left(1 - \frac{r^2}{a^2}\right) - \frac{E\zeta}{\epsilon_r\epsilon_o\eta}\left[1 - \frac{I_o(\kappa r)}{I_o(\kappa a)}\right]. \tag{2.25}$$

Equation 2.25 shows the behavior of many operational conditions that can be performed. They are all plotted in Figure 2.18. Figure 2.18A shows the velocity profile when only EOF is present. Note again that this velocity is very different to the parabolic (or Poiseuille) flow caused by pressure gradients. If $\kappa a > 100$, which is the case in most ESTs, then the EOF velocity is constant along the entire cross section except for very close to the wall where it falls from $v_{eo} = -\frac{\zeta E}{\epsilon_r\epsilon_o\eta}$ to zero. Figure 2.18B shows the velocity profile of a pressure driven flow (Figure 2.13 shows the same as contour levels). In Figure 2.18C EOF acts in the same direction as pressure driven flow. Figure 2.18D shows a very odd situation in which a pressure driven flow acts against the EOF. The flows of Figures 2.18C and 2.18D are produced, for instance, when the EST is operated with unleveled reservoirs.

2.9.2 EOF in Rectangular Microchannels

Square or rectangular microchannels filled with a liquid and $\kappa a \gg 100$ are subjected to an EOF that is flat in the whole liquid phase, with exception only very close to the walls where it falls to zero. In this case, using equation 2.23 again, the maximum velocity and volumetric flow rate are given by:

$$v_{max} = \frac{\mu_{eo}V}{L} \quad \text{and} \quad \frac{\Delta \text{Vol}}{\Delta t} = \frac{4\mu_{eo}V}{L}wh, \tag{2.26}$$

where h is half the height of the microchannel and w is half the width, so that the cross sectional area is given by $A = 4wh$. Recall once again that all results in this section refer to flows with low Reynolds number (laminar flow).

2.10 Supression of EOF

The observed electroosmotic mobility variations, both from run-to-run and day-to-day, have always been problematic, especially for capillary and microchip platforms. The variations come from the chemical and physical changes that take place at the inner solid–liquid interface. The chemical components of the successive injected sample plugs and their respective analytes promote chemical and physical changes to the solid wall (diffusion of ions into the dielectric solid), Stern layer (changes in thickens and constitution), and the shear plane (viscosity changes). Even worse, a hysteresis effect has been noticed [24]. All of these phenomena change the ζ potential (electrokinetic potential) and the viscosity at the shear plane from run-to-run, and this leads to changes in electroosmotic mobility and velocity. These variations promote undesired uncertainties in migration times and errors in quantitative analyses, as peak heights are affected by migration times (they decrease with time) and peak widths (and consequently areas) are affected by their apparent velocities when they cross the detection point (slower velocities give wider peaks). *Electroosmosis has always been a rebel variable!*

Nevertheless, there are some advantages to the presence of an EOF in the open layouts, which are the techniques that possess an inlet for the sample, a crossing line for the analytes at the detection point and an outlet for the analytes. The three advantages are: (1) anions, cations, and neutral components can be analyzed simultaneously from a single injection into the inlet, as the EOF drives them all to the detection point; (2) some analytes are easy to separate and therefore their run times can be shortened. This can be achieved by driving the electroosmotic flow from the inlet to the outlet, without the need to change the capillary; and (3) very high separation resolutions can be achieved by using an EOF that is reversed to run against the analytes, with a velocity that is fine-tuned to be a bit smaller or larger than the electrophoretic mobility of the analytes. In this case the analytes perform very long runs in the "reference frame of the liquid phase", which gives very high resolutions but maintains the inherently poor repeatabilities of EOF.

2.10.1 Protocols for EOF Suppression

A suppressed EOF is advantageous in many situations (they are discussed in Section. 2.10.2). There are four ways to suppress the EOF. The simplest is to lower the pH of the running buffer solution, which will neutralize the silanol groups at the quartz and glass walls, reducing the ζ potential of equation 2.23 to close to zero. However, this restricts the pH ranges of the buffer solutions that can be used and a residual and unstable EOF may persist.

The second way to suppress the EOF is to use *non-covalent coatings*, they are also called dynamic coatings. This is based on the use of small molecules (alyphatic polyamines), cationic surfactants (cetyltrimethylammonium bromide and dodecyltrimethylammonium bromide) or polymers dissolved in the rinsing buffer. They self-attach to the surface of the inner wall of bare microchannels and capillaries [25]. Cationic surfactants are able to reverse the EOF, however they have multiple effects on the separation process that are hard to predict, as they depend on the concentration used, composition of the BGE, temperature, and nature of the analytes under analysis (size, charge, and valence). Neutral polymers are very popular and they attach to the fused silica and glass bare walls because they are slightly hydrophobic and have a higher affinity to the bare wall than to the bulk BGE. Examples of neutral polymers that self-attach to the wall includes poly(vinylalcohol), poly(ethylene oxide), poly(vinylpyrrolidone), poly(dimethylacrylamide), and cellulose derivatives, among others. The adsorption creates a high viscosity zone across the shear plane, which severely suppresses the EOF regardless of the remaining *zeta* potential. In this case the EOF suppression is better understood with equation 2.27, which was proposed by Hjertén [26].

A third and very reliable procedure to suppress the EOF is the use of *covalent coatings* [27], also called permanent coatings or covalent modifiers. Here, the polymers are chemically bound to the dielectric material of the platform wall, providing two benefits: less ionizable silanol groups will remain following the reaction and the polymers will persist for longer despite the inter-run rinsing steps. A great variety of polymers can be bound to the fused silica and glass walls. The protocols for polymerizing linear acrylamide starting from the wall, developed by the pioneering work of Hjertén and collaborators, are reliable and show a good cost-benefit ratio [28]. Many more methods have been published over the years, including the covalent bonding of nanoparticles to the wall (Hajba and Guttman [27] provided an updated list of non-covalent and covalent coating protocols).

Finally, the fourth is based on a layer of cross-linked polymers that is produced on column and physically attached to the wall using a thermal treatment [29,30]. These protocols require the handling of viscous solutions, high pressure, and thermal treatments, which makes them difficult to perform in most of the application oriented laboratories. However, the resulting permanent coatings have proven to exhibit good performances that last for hundreds of runs.

The calculation of μ_{eo} when polymers are localized across the shear plane is given by:

$$\mu_{eo} = \frac{1}{\epsilon_r \epsilon_o} \int_0^\zeta \frac{1}{\eta} d\psi \,, \tag{2.27}$$

where η is small in the bulk of the liquid phase (where $\psi = 0$) and very high close to and at the shear plane, where $\psi = \zeta$.

2.10.2 Advantages of Suppressing EOF with Covalent Coatings

It is well known that covalent coatings diminish the adsorption of macromolecules onto the wall, especially the absorption of proteins. Coated microchannels and capillaries also give narrow and symmetrically shaped peaks, leading to higher separation efficiencies. The first explanation for this is that the polymers covalently bonded to the silanol group reduces the surface density of charges at the wall, and this reduces the probability of adsorption of macromolecules to the wall. The second factor is that, even with charges on the wall, the steric effect that arises from the presence of the polymers bound to the wall prevents the proteins and other macromolecules to be adsorbed by the layer of charges.

In summary, covalent coating prevents the absorption of macromolecules onto the walls and subsequent poor resolutions. Moreover, a reliably suppressed EOF brings high repeatabilities and reproducibilities to both migration times and quantitative analyses. However, together with a good coating procedure, a good rinsing protocol is also required for these coatings to be stable and give reproducible quantitative analysis for a large number of runs, for instance more than two hundred runs [31].

There are some advantages in operating the toroidal layouts with a suppressed EOF. As shown in Chapter 4, cations and anions can be analyzed simultaneously without the need of an EOF. This is made by opposite dual-injection of the sample. In this case both anions and cations run in a clockwise manner. The analytes cross the detection point(s) many times, starting just after the injection event and running until the desired resolutions are achieved. In other words, the progress of the separation can be monitored at given time intervals. Moreover, higher resolutions are possible in these toroidal layouts (in the capillary, microchip or slab platforms) in a reproducible and repeatable manner because EOF is suppressed.

2.10.3 Measuring Small and Large EOF Velocities

Many methods have been proposed for the measurement of electroosmotic velocities and mobilities [32–38]. Knowledge of these parameters is very important for many applications, including evaluation of the quality of a non-covalent or covalent coating. A method introduced by Zhang et al., [39], called the "two-step method", is based on the use of a neutral marker that is injected into the capillary and run twice. A small and precisely constant pressure is applied to both runs, but the polarity of the high voltage is reversed between runs. If the difference between the migration times of the runs is null, then $\mu_{eo} = 0$. However, the presence of a small EOF is revealed by a difference in the migration times t_1 and t_2. This is quantified by equation 2.28, which works for both small and large EOFs. The same method is also expected to work for a microchannel in a chip.

The "two-step" method for μ_{eo} measurement is:

$$\mu_{eo} = \frac{|t_2 - t_1| L\, l}{2\, t_1\, t_2 V}\,, \tag{2.28}$$

where t_1 and t_2 are the migration times, and L and l are, respectively, the total and effective lengths.

2.11 Joule Effect and Heat Dissipation

The tendency of a solution to allow the flow of ions as a consequence of the application of an electric potential difference is measured by the physical quantity called conductivity. Inversely, the difficulty a solution imposes to the flow of the current is called resistivity. In the so-called Ohmic regime the current that flows through the solution is proportional to the applied electric potential difference and is inversely proportional to resistivity. However, unlike electrons in metals, the conductivity of ions in liquids increases with temperature. This means that resistivity decreases with an increase in temperature.

An ion or charged particle with charge q that is placed in a vacuum under the influence of the electric field E accelerates and acquires kinetic energy. This acceleration is given by $a = qE/m$, where m is the mass of the ion. In liquids, however, the ion or charged particle accelerates for a fraction of a nanosecond only and then collides with a surrounding molecule or ion, converting its kinetic energy to heat. These repeated events give the ion or charged particle an apparent constant average velocity but increase the temperature of the solution due to the collisions. The heat generated by the passage of current is the called *Joule effect* (in recognition to the British physicist James Prescott Joule, 1818–1889. The unit of energy, Joule, was named in his honor) and the rate of heat generated per unit volume is proportional to the number of ions per unit volume (concentration), the electrophoretic mobility of the ions (μ_e), and the applied electric field strength (E). At the same time heat also dissipates (diffuses) and is removed from the external wall by stagnant air, forced air, circulating liquid coolant, or a thermoelectric cooled surface (based on the Peltier effect). As a result, temperature profiles build up in the separation media and in the surrounding dielectric material used to build the platforms. These temperature profiles are studied in Section 2.12. Therefore, buffer solutions with the highest possible buffer capacity (see Section 1.1.12) and lowest possible electrical conductivity, but still compatible with the sample's conductivity, are the most recommended. Heat removal strategies are very important to all ESTs. There are also a few modes of operation (constant current, power or voltage) that also deserve some attention.

Certain high voltage power supplies can be set to run at a constant current, power or voltage. This promotes the migration of the analytes as well as the BGE's anions and cations, and as a consequence heat is always generated because of the Joule effect. The relationship between power and current is given by equation 2.29 and the relation between power and voltage is given by equation 2.30. The three operation modes mentioned above have both advantages and disadvantages. From equation 2.30 we can see that the application of a constant power (P) results in a lower voltage when the resistance (R) decreases (when temperature increases due to Joule heating). The application of a constant current also has at least one advantage: if temperature increases then resistance (R), and consequently dissipated power, decreases. The opposite happens when a constant voltage (or constant electric field) is applied: when temperature starts to rise the resistance decreases, making the dissipated power increase, which in turn increases the temperature even further. The advantage of this mode of operation is that electric field strength is constant throughout the run, which allows the calculation of electrical mobilities using the simple equation $\mu = v/E$. However, large temperature differences may build up at the center of the separation phase, causing band broadening (see Section 2.13). This limits the electrostatic potential differences that can be applied to all ESTs.

The dependence of power (P) on the resistance (R) and current (i) is given by:

$$P = R i^2. \tag{2.29}$$

The dependence of power (P) on the applied electric potential difference (V) and resistance (R) is given as:

$$P = \frac{V^2}{R}. \tag{2.30}$$

2.12 Temperature Profiles

ESTs are based on the application of an electric field, which drives the analytes along their separation path and also drives the ions of the BGE. Therefore, a constant quantity of heat is continuously generated per unit volume of the separation medium. Temperature gradients form at the start of the runs when the high voltages are first turned on. In a short amount of time these temperature profiles will reach an equilibrium due to the balance between heat generation and heat removal.

Temperature affects the conductivity of the buffer solutions (it increases with temperature), the viscosity of the solutions (it decreases with temperature), the mobility of the ions (it increases with temperature), and the separation efficiencies of the ESTs in many ways. For instance, the pK_a of some amines changes significantly with temperature, which may affect the degree of ionization of these analytes when the buffer pH does not also change with temperature in the same manner. Conversely, the pH of some buffers (e.g. TRIS) changes with temperature, which in turn changes the average charge (degree of ionization) of some analytes whose pK_a values are less affected by temperature or have an opposite sensitivity to temperature.

The approximate dependence of electrophoretic mobility on temperature is given by:

$$\mu_e(T) = \mu_e(25\,^\circ C)[1 + k(T - 25\,^\circ C)], \tag{2.31}$$

where k is an empirical parameter that depends on the analyte and buffer solution used. In aqueous dilute solutions $k \simeq 0.025/^\circ C$ in the 20 to 30 °C range.

There are many heat removal strategies for capillaries, starting from stagnant air and progressing to forced air and forced liquid coolants. Forced liquid coolants assure an almost constant temperature over the external surface of the polyimide (air–polyimide interface). In this case heat is generated in the capillary core, diffuses through the fused silica wall, through the polyimide coating, and into the circulating coolant (air or liquid). The heat diffusion equation in cylindrical coordinates with a source of heat term must be solved to find the temperature profiles for this case. This is given by equation 2.32.

The heat diffusion equation with a heat source term:

$$\frac{1}{r}\frac{\partial}{\partial r}\left(r\frac{\partial T}{\partial r}\right) = -\frac{p}{\lambda}, \tag{2.32}$$

where λ is the thermal conductivity coefficient of the liquid and p is the rate of heat generated per unit volume in the liquid phase due to the Joule effect (the SI unit of p is Joule $m^{-3}\ s^{-1}$). Note that $p = \rho j^2$, where ρ is the resistivity of the liquid phase and j is the applied density current. The parameter j is related to the total current by $i = \pi a^2 j$, but only if the density current is constant over all points of a cross section of the liquid phase.

Equation 2.32 is identical to 2.12, the only difference being the non-homogeneous term. This has important practical and theoretical implications that are explored in Appendix F.

The solution of the heat diffusion equation in cylindrical coordinates with a heat source term (equation 2.32) produces the following simple solution:

$$T(r) = T_a + \frac{pa^2}{4\lambda}\left(1 - \frac{r^2}{a^2}\right) \quad \text{with} \quad 0 < r < a, \tag{2.33}$$

where $r = a$ is the position at the liquid–fused silica interface or capillary inner radius and T_a is the temperature at the inner wall of the capillary. However, T_a depends on the rate of heat dissipation through the fused-silica wall. This can be found by solving equation 2.32 again, but now with the border condition of a heat flux at $r = a$. This gives:

$$T(r) = T_b + \frac{pa^2}{2\lambda_s}\ln\left(\frac{b}{r}\right) \quad \text{with} \quad a < r < b, \tag{2.34}$$

where λ_s is the thermal conductivity coefficient for the fused silica, $r = b$ is the position at the fused silica-polyimide interface and T_b is the temperature at this fused silica-polyimide interface. Note that $T_a = T_b + [pa^2/(2\lambda_s)]\ln(b/a)$. Now T_b must be calculated for T_a to be known. The total heat flux at $r = b$ is identical as at $r = a$. Therefore:

$$T(r) = T_c + \frac{pa^2}{2\lambda_p} \ln\left(\frac{c}{r}\right) \quad \text{with} \quad b < r < c, \tag{2.35}$$

where λ_p is the heat diffusion coefficient of the polyimide and $r = c$ is the position of polyimide–coolant interface. Supposing that the coolant is very efficient in removing the heat from the polyimide surface and that the temperature is constant and equal to T_c over all this surface. Therefore, the final expression of the temperature inside the liquid phase, which is the most important property to know, is given by:

$$T(r) = T_c + \frac{pa^2}{2\lambda_p} \ln\left(\frac{c}{b}\right) + \frac{pa^2}{2\lambda_s} \ln\left(\frac{b}{a}\right) + \frac{pa^2}{4\lambda} \left(1 - \frac{r^2}{a^2}\right) \quad \text{with} \quad 0 < r < a. \tag{2.36}$$

T_c is considered as both the temperature of the coolant and the temperature at $r = c$, which is the polyimide–coolant interface. In the great majority of practical applications this is not exactly the case as the coolant always hits one side of the capillary and a vortex with a higher temperature (less efficient heat removal) is observed on the opposite side. Figure 2.19 illustrates the ideal (blue curve) and more realistic (red curve) scenario.

In conclusion, forced air or a dielectric circulating coolant is recommended to stabilize the temperature profiles across the dielectric materials used in the platforms and, more importantly, in the liquid phase were the heat is generated. To avoid large temperature differences between the coolant fluid and the separation medium, support materials made of good thermal conductors are recommended, this can be seen in equation 2.36 and Figure 2.19. However, good electric insulators are, in general, not good thermal conductors, and this is still an active field of research in material sciences. The distance between the circulating coolant and the heat source (separation medium) is also very important, and should be as close as possible, but not too close to avoid disturbances on the ζ potential due to leakage of electric current to the coolant stream.

To solve the heat diffusion equation with a source term for the microchip and slab platforms, then Cartesian coordinates are recommended. In these cases the recommendation is to expressed the solution as the sum of cosines, this gives simple results and similar to that of equation 2.19. The knowledge of these temperature gradients leads to novel setups, discussed in Appendix F.

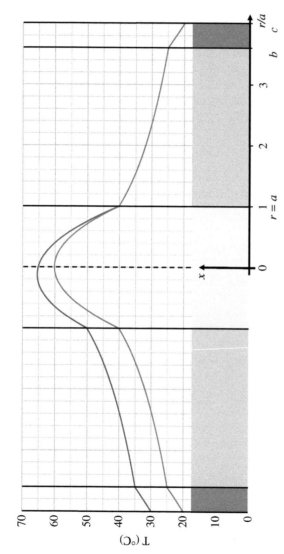

Figure 2.19 Radial temperature profiles of a capillary that has 180 μm OD, 50 μm ID and 10 μm CT (polyimide external coating, see Figure 2.3). The liquid–fused silica interface is located at $r = a$, the fused silica–polyimide interface is at $r = b$, and the polyimide–coolant interface at $r = c$. The blue line shows an ideal situation in which the whole polyimide–coolant interface is set to 20 °C. The red curve shows a more realistic situation in which the forced air hits the right side of the capillary. Heat removal is efficient on this side and less efficient on the opposite side due to the air vortices formed, leading to asymmetrical temperature profiles with respect to the central axial direction (dashed line).

2.13 Molecular Diffusion and Band Broadening

Molecular diffusion is considered an unavoidable mechanism of band broadening in liquids and is represented by the diffusion coefficient D (also known as the self-diffusion coefficient). It is a consequence of the translation degrees of freedom of the ions and molecules in liquids. This dispersion is promoted at the expense of thermal energy. The Brownian motion of microscopic particles, which can be seen under a microscope, is a consequence of this phenomenon. There is a relationship between diffusion coefficient and temperature that is valid in the limit of low Reynolds numbers and was independently found by William Sutherland in 1905 [40], Albert Einstein in 1905 [41], and by Marian Smoluchowski in 1906 [42]

$$D = \frac{k_B T}{f} = \frac{\mu_e k_B T}{q},$$ (2.37)

where k_B represents the Boltzmann constant, T represents temperature in Kelvin, f is the friction coefficient of the molecule, μ_e its electrophoretic mobility, and q the charge. The right hand side term of equation 2.37, also known as the Nernst–Einstein equation [43,44], is expected to be valid only when the frictional coefficients involved in electrophoretic transport are identical to the frictional coefficients involved in hydrodynamic transport. This is not the case, for instance, for polyanions and polycations, as demonstrated for DNA in free solution [45].

The equation that governs molecular diffusion can be deduced by applying the continuity equation to Fick's first law (see Appendix E). For homogeneous and isotropic environments, which are usually observed in liquid solutions, the diffusion equation for one spatial dimension is given by equation 2.38. The concentration c is shown to decrease very quickly ($\partial c/\partial t \ll 0$) at the tips of sharp bands, because the second spatial derivative is very large and negative at these positions. The opposite happens in sharp valleys. The only positions where the concentrations stay unchanged are at the inflection points of the bands, where $\partial^2 c/\partial x^2 = 0$.

The diffusion equation in one spatial dimension:

$$\frac{\partial c}{\partial t} = D \frac{\partial^2 c}{\partial x^2},$$ (2.38)

where D is the molecular diffusion coefficient, assumed to be time independent.

According to the diffusion equation, the variance of a molecular distribution grows in time as $\sigma^2 = 2Dt$ and the standard deviation as $\sigma = (2Dt)^{1/2}$ when only molecular diffusion is taken into account. However, it is common to note a few additional band broadening mechanisms in action in real practical situations. This comes from many sources and they all have a detrimental effect on separation efficiency: axially inhomogeneous EOF [46], BGE inhomogeneities [47], capillary coiling [48], injected plug length, parabolic laminar flow due to pressure differences or unleveled reservoirs [49], radial temperature gradients [50], and wall adsorption [51,52], to mention only a few.

Sometimes, the total variance can be calculated as the sum of all contributions. However, the additivity of the variances is only possible if the dispersion sources are independent of each other. The presence of non-linear effects, for instance electric field distortions caused by the high conductivity of an analyte band, produces an extra dispersion mechanism that depends on the length of the injected sample and on the variances of the other sources. However, the additive rule is, in general, valid when the analytes are at low concentrations, the buffer has a similar conductivity to the sample, and the BGE constituents have mobilities similar to the analytes. In this case *dispersion coefficients* (D_i) are used for the time dependent variances and their contributions are also calculated as $\sigma_i^2 = 2D_i t$. The total variance is then easily calculated as:

$$\sigma^2(t) = \sigma_{inj}^2 + \sigma_{det}^2 + 2Dt + 2D_{ads}t + \dots$$
$$+ 2D_{coil}t + 2D_{emd}t + 2D_{ieof}t + 2D_{pvf}t + 2D_{temp}t + \dots . \quad (2.39)$$

The first line contains time independent contributions to the total variance, namely, the variance of the initial injected plug and the contribution from the detection width. There are other time independent variances and these are discussed in Chapter 4. Note that the term σ_{det}^2 affects only the peak widths in the electropherograms and does not affect the bands themselves. Hence the importance of distinguishing bands from peaks (see Section 2.6).

There is a large number of time dependent contributions to the total variance, such as: molecular diffusion (D), wall adsorption (D_{ads}), capillary coiling (D_{coil}), electromigration dispersion or BGE inhomogeneities (D_{emd}), parabolic velocity flow driven by pressure or unleveled reservoirs (D_{pvf}), inhomogeneities in the EOF (D_{ieof}), and temperature gradients (D_{temp}), to cite a few.

Mathematical equations have been found over the years relating the dispersion coefficients to relevant variables. The time dependent dispersion coefficient that is related to the parabolic laminar velocity profile [49], for instance, is given by:

$$D_{pvf} = \frac{a^2\,\bar{v}^2}{48D} = \frac{a^2\,v_{max}^2}{192D}, \quad (2.40)$$

where a is the inner radius of the capillary, $\bar{v}$ is the average fluid velocity, and D is the diffusion coefficient. Note that for a cylinder $\bar{v} = v_{max}/2$, where v_{max} is the maximum velocity given by equation 2.14). These parabolic flows may occur by mistake (unleveled reservoirs) or intentionally in the so-called *pressure assisted capillary electrophoresis*. In this case the separations are accelerated or decelerated, by an external pressure driven flow, either with or against the direction of EOF.

The variance due to an initially injected rectangular band with length b is given by:

$$\sigma_{inj}^2 = \frac{l_{inj}^2}{12}. \tag{2.41}$$

The variance added to the peaks in the electropherogram due to the detection length l_{det} is given by:

$$\sigma_{det}^2 = \frac{l_{det}^2}{12}. \tag{2.42}$$

In this case it is assumed that the detection sensitivity is constant along the length l_{det} (rectangular function) and zero outside. These peak broadening can be removed from the electropherograms using *deconvolution* (an algorithm used to reverse the effects of convolution on recorded data) and it works for any sensitivity function in the interval of length l_{det}, provided that this sensitivity function is known. This enlargement effect is common in absorption detectors and contactless conductivity detectors, but are negligible in laser-induced fluorescence detectors (used in capillary and microchip electrophoresis) and UV based fluorescence detections used in transilluminators (for slab electrophoresis).

The Table 2.2 gives the band broadening mechanisms listed in alphabetic order and some preventive measures that can be taken.

This list in Table 2.2 should be adopted with the goal of improving the separation efficiency and/or resolution (see Section 3.5). With the above precautions it is easy to mitigate all band broadening mechanisms, with only molecular diffusion remaining as the predominant band broadening mechanism. This can be proven using forth and back electrophoresis and plotting the variance of the peaks as a function of the crossing times [53]. If the slope of this line is higher than D, the molecular diffusion of the analyte, then other band broadening mechanisms are present beside molecular diffusion.

Band broadening mechanisms affect the real objects called spatial molecular distributions or bands. The band shape of a given analyte (i) is given by the distributions $c_i(x, t)$, see Section 2.6. However, as mentioned before, there are "peak" broadening mechanisms that affect only the peaks in the electropherograms and not the bands themselves. They are listed in Table 2.3 separately to emphasize the distinction between bands and peaks.

2.14 Sample Stacking and Band Compression

There are dozens of *sample stacking* methods that have been developed over the years, which are applicable to many separation modes. They are used to narrow the sample plug length at the start of the runs, producing two benefits:

Table 2.2 Troubleshooting checklist to reduce band dispersion in ESTs.

Band broadening	Preventive measures
Axially inhomogeneous EOF	Better flushing between runs
	Use non-covalent coatings
	Use covalent coatings
	Clean the sample
BGE inhomogeneity	Adopt a sample cleaning step
	Use BGE with similar mobilities of analytes
	Dilute the sample (avoid overloading)
Capillary or channel coiling	Adopt a narrower microchannel
	Adopt a capillary with smaller ID
Injected plug length	Inject less sample
	Lower the sample's conductivity
	Use a sample stacking method
	Use band compression if feasible
Molecular diffusion	Lower the running temperature
	Increase the buffer viscosity
	Derivatized analytes exhibit lower D
Parabolic laminar flow	Check for unleveled reservoirs
	Check for pressure differences
	Check for clogged gas tubes
	Check electrolysis gas relief
Temperature gradients	Apply lower field strengths
	Use buffers with lower conductivity
	Ensure efficient heat removal across all parts
	Consider using zwitterionic buffers
Wall adsorption	Better flushing between runs
	Use non-covalent coatings
	Use covalent coatings
	Change the pH

Table 2.3 Procedures to reduce peak width caused by factors other than band dispersion.

Peak broadening	Preventive measures
Detection length	Shorten the detection length
	Apply the deconvolution operation
Very long transit times	Increase field strength (E)
	Use coordinate transformation
	EOF modulation with axial fields
	Consider using a spatial scanner

1. Higher resolution or, alternatively, a decrease in the run time necessary to achieve satisfactory separations, as the initial variance will be small compared to the variance added during the run by molecular diffusion;
2. Increase the detection limit of the analytes. The inner diameter of the liquid core in capillaries and the heights and widths of microchannels in microchips are usually in the 25 μm to 100 μm range. This brings difficulties to the detection systems as only very small volumes are available for detection.

Figure 2.20 illustrates a stacking event: at the beginning of the run a long sample plug is injected. Some phenomena are intentionally and rationally used to stack this long plug into a narrow band. Fortunately there is a series of meticulous review articles reporting the development of these stacking methods over the years and the respective theoretical models used to describe them [55–60].

There is one unavoidable stacking or unstacking event that always takes place at the start of the runs in many separation modes (AFE, ELFSE, FSE, MEEKC, MEKC, and SE) and is related to the ratio of sample plug to buffer conductivity. Under ideal conditions, if the sample plug has a conductivity lower than the buffer solution used in the separation medium, then the sample will be subjected to a stacking or narrowing event. This is called an *intrinsic sample stacking* procedure, which occurs because the field inside the plug is amplified due to the lower conductivity. A similar event is produced in slabs by the use of a stacking gel and a running gel. The stacking gel is made by using a running buffer that is diluted 5 to 10 times. This narrows the bands' standard deviation by a factor of 5 to 10 when they cross the interface of the stacking gel into the running gel.

Conversely, when the conductivity inside the plug is higher than the conductivity outside, the sample will be dispersed and leave the injected plug zone much wider. This is one reason why conductive buffers (solutions with electrolytes) must be used for the above mentioned modes and others. Consequently, the Joule effect

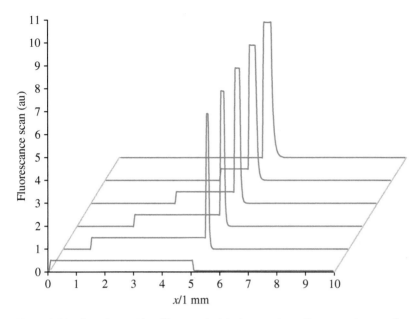

Figure 2.20 Sample stacking illustrated with six snapshots. There are dozens of different stacking procedures that are performed at the inlet (beginning of the runs) in order to improve the limits of detection, especially when absorption detection is used. The injected analyte plug lengths are reduced by a factor of approximately ten and the peak heights are increased by a similar factor.

and the subsequent radial temperature gradients produced will always be present in most ESTs.

Stacking is a term reserved for these events which are performed at the beginning of the runs and which aim to produce narrow bands at the start of the separation process. *Band compression*, on the other hand, is performed on-line during the runs, with the objective of compressing target bands [61,62]. In the original work it was called stacking, however the appropriate name of this is band compression as it can be performed on-column and anywhere, except at the inlet because the bands must slow down while entering a selected segment of the microchannel or capillary.

There are at least two simple ways to achieve this:

1. Heating or cooling a segment of the separation medium just before the target bands start to enter this specified segment. The velocity of the target analytes slows down and a piling-up event takes place. For this to happen a pH sensitive buffer must be used when analytes with a temperature insensitive pK_a are separated [61]. Conversely, a temperature insensitive buffer must be used when analytes with a temperature sensitive pK_a are under analysis [62]. In

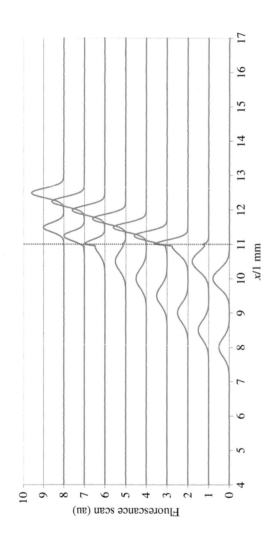

Figure 2.21 Nine snapshots illustrating an on-column band compression event. The vertical red lines show the segment of the separation medium that was chosen to promote the band compression. This is accomplished, for instance, by using an independent temperature control on this segment and a temperature sensitive buffer [61]. The pH of the buffer changes locally when the temperature of this independent thermal zone is changed, and this in turn changes the charge and mobilities of some analytes when they enter this zone. Once the bands of interest are all inside this thermal zone then its temperature is reversed back to the normal running temperature. The opposite also works: modulating the pK_a of temperature-sensitive analytes that are run in a temperature insensitive buffer [62]. These band compression events are performed during the runs with the purpose of reducing band variance and, consequently, increasing peak capacity and the limits of detection. The underlying phenomena of band compression are distinct from the underlying phenomena of stacking procedures.

summary, buffer solutions and analytes with opposite temperature sensitivities are necessary to allow the mobility of the analytes to be modulated through temperature modulation. Table 7.2 gives the dpK_a/dT of a list of compounds that are commonly used to prepared buffer solutions.

2. Shining light at a segment of the separation buffer that contains a *photoacid* or *photobase* just before the entrance of the target analytes into this segment. Some organic molecules are weak bases (or weak acids) in the dark. However, when they are excited by a certain wavelength, they become strong bases (or acids) and their pK_a may change by five to seven units. This dramatically changes the degree of ionization of the analytes and their mobility which have a pH dependent net charge.

Figure 2.21 illustrates a band compression event that is performed on-column, usually far from inlet and outlet because some space is required for the hardware (independent temperature zone [61,62] or light incidence zone). These band compression events do not significantly improve the band-to-band resolution because the distance between bands is reduced by the same factor as the standard deviations of the bands. However, narrower bands improve peak capacity and limits of detection, and prevent the bands from vanishing into the baseline after extended run times, as it may happen in toroidal electrophoresis of hard to separate mixtures. These compression events have interesting applications in the toroidal layout, especially when applied in a cyclic manner (see Appendix G).

2.15 Separation Modes

2.15.1 Affinity Electrophoresis

Affinity electrophoresis (AE) is a collective term for ESTs that make use of a substance (ligand or selector) in the buffer that is capable of establishing an equilibrium complexation reaction with the analytes. Simply put, the ligand interacts selectively with the components of the sample to optimize the separation of the analytes from each other and from the sample interferents. These interacting entities (selectors) are added to the buffer solution, creating a true solution, or are covalently bound to an immobile gel, wall, or free polymers. Analytes with identical electrical mobility (in FSE, MECK, and many other separation modes) can be separated due to the differences in the ligation equilibrium constants, which affects the effective electrical mobility. In this case the separation mode is a hybrid of AE and FSE, or AE and MECK, and so on.

Examples of such ligands are antibodies for the separation of antigens, or vice versa. Aptamers, also known as synthetic antibodies, also have an interesting potential as they interact with endogenous substances, against which it is difficult to produce normal antibodies. Moreover, aptamers are more stable and easy to replicate. Molecules able to form inclusion complexes are also an important class of ligands with important applications. Examples of such molecules are cyclodextrin, sulfated-cyclodextrin, and crown ethers. The inclusion complexes of pharmaceutical compounds and L- and D-amino acids into beta-cyclodextrin, for example, is very popular in the quantitative separation and analyses of mixtures of stereoiomers. In this case these additives work as chiral selectors.

Another class of ligands are the chelating agents such α-hydroxyisobutyric acid, valinomycin, 2,6-pyridinedicarboxylic acid, and others. They are very important for the separation of small ions. In this a complexation equilibrium is established with the mono or polydentated ligand and the cations (usually transition metals). This is used to improve the separation efficiency and resolution among the ions. Some chelates of EDTA with cations end up forming very stable anionic complexes. They are used for the simultaneous analysis of anions and cations (as anionic complexes). A detailed study and updated list of targets and ligands is given by Yu et al [63].

Considering an analyte that migrates with its own electrical mobility μ_A and an analyte-selector complex that migrates with a mobility μ_{AS}, then in the 1:1 complexation stoichiometry, in which the selector concentration c_S is not affected by the concentration of the other analytes, the following are obtained [64]:

$$\mu = \frac{\mu_A + K'_{AS} \, c_S \, \mu_{AS}}{1 + K'_{AS} \, c_S} \quad \text{or} \quad K'_{AS} = \frac{\mu_A - \mu_{AS}}{\mu - \mu_A}, \tag{2.43}$$

where K'_{AS} is the apparent binding constant (in contrast to the true thermodynamic). Equation 2.43 can be used either to predict μ if K'_{AS} is known, or to determine K'_{AS} by measuring μ. Note that μ_A is measured by simple using $c_S = 0$. For the measurement of μ_{AS} then c_S must be high enough to assure that the complexation reaction is displaced towards its completion.

The migration time shifts between runs, with and without the complexing agent (ligand or selector), are also used for the determination of the complexation constants. The ESTs are also used for the determination of the binding constants themselves, which is the opposite of what was discussed in the previous paragraphs, and they are very competitive compared to the other techniques (cryo-electron microscopy, fluorescence spectroscopy, isothermal titration calorimetry, nuclear magnetic resonance, surface plasmon resonance, among others). Capillary electrophoresis, for instance, can be used to measure the affinity constants at the level of small molecules, macromolecules, single cells, between cells, and probably among unicellular organisms [63]. This is unrivaled compared to the other technologies.

The AE separation mode should not be confused with MEEKC or MEKC, which make use of a distinct phase (microemulsion or micelle, respectively) where the analytes have the freedom to diffuse into. The underlying separation mechanism in this cases is the difference between the distribution coefficients of the analytes between the hydrophilic (hydrophobic) phase and pseudophase formed.

2.15.2 Electrochromatography

Electrochromatography (EC) is a type of liquid chomatography that uses electroosmotic pumping to make the mobile phase flow instead of using conventional pressure driven only. In fact, only electroosmotic pumping is sufficient to drive the mobile phase in some operations, only pressure in others, while some use a combination of both. The platform of preference is a capillary made of glass or fused silica with an ID smaller than 200 μm, affording higher field strengths. They may be operated as packed or open tubular. The potential advantages of EC were perceived a long time ago [65] and demonstrated on the second part of one of the papers of 1981 from Jorgenson and Lukacs [66]. These advantages are:

1. The flow characteristics (Poiseuille) of the mobile phases used in all chromatographies promotes a significant band dispersion, which limits the number of theoretical plates, the resolutions, and peak capacities that can be achieved. The EOF has an ideal profile in this regard, as it causes no extra band dispersion.
2. It is easy to keep a constant EOF rate in columns with a small ID, irrespective of packing and length. No high back pressure is observed in the systems. According to equation 2.23, the rate of EOF will stay approximately the same as long a constant field strength (Volts/centimeter), temperature, packing, and buffer solution are used.
3. The *separation selectivity* [67] can be improved, since the analytes are subjected to both electrodriven migration and distribution between the mobile and stationary phases.

However, some difficulties arise when it comes to prototyping and operating these systems. Pressure driven flows are periodically applied for column cleaning and regeneration. Flushing the columns with *column volumes* in the 5 to 20 range with the appropriate solvent, depending on the stationary phase and packing material used, are routinely applied in chromatography. However, high back pressures are predicted for columns that are good for EC, and even higher back pressures are observed in contaminated columns. This in turn requires tight connections on the fused silica or glass column inlet to support these high pressures. Note that the high-voltage used must be also compatible with these connectors and detection system used.

Additionally, gradient elution is required for the separation of large molecules, but this is not easy to implement in the EC instruments as the high voltages need to by applied either at the inlet or outlet. Proteins are notorious for adsorbing to the stationary phases used in chromatography and this brings additional difficulties to both the separations and the column regeneration process. Finally, the EOF is usually small in non-aqueous solutions, therefore aqueous solutions must be used in order to produce small retention times and to afford the benefits of the minimal band dispersion caused by the EOF. This in turn limits the applicability of electrochromatography, as the solutes must be soluble in aqueous solutions, unlike MEKC and MEEKC, which exhibit good separation efficiencies for neutral analytes. The micelle-in-oil and microemulsions of water-in-oil are suitable for neutral compounds that are not soluble in water.

In summary, from the cost-to-benefit ratio perspective, it is safe to say that the costs (difficulties) of electrochromatography are high, but so are the potential benefits (high separation efficiency). There are hundreds of successful applications to prove these benefits [68,69] and notably in the open-tubular format of capillary electrochromatography [70].

2.15.3 End-labeled Free-solution Electrophoresis

All linear polyanions and polycations have a total charge that is proportional to their number of monomers (length) and total friction coefficient that is also proportional to their number of monomers because they migrate as open and free draining chains. This subtle but unfortunate property makes it impossible to separate them in free-solution electrophoresis because the charge to friction coefficient is approximately same for all lengths (oligomers).

However, there is an approach that consists of binding a neutral drag tag to one or both of their ends in order to break this charge to friction coefficient balance, so that they can be separated by length in free-solution electrophoresis, this was named end-labeled free-solution electrophoresis (ELFSE) [71]. The first structures separated by length using ELFSE were highly charged oligosaccharides [72] and heparins [73]. Soon after, attention was turned to DNA sequencing. In this case the drag tag is very important (large and monodisperse) to achieve very long read lengths. Initially the biotin-streptavidin system was used and more than one hundred bases were sequenced (Figure 2.23) [74] and many others where successfully applied: polypeptoids [75], genetically engineered protein polymers [76,77], comb-like monodisperse polypeptoids [78], and 502 bases were sequenced using worm-like micelles as the drag tag [79].

The mobilities in these cases are easily calculated for RNA, single strand DNA (ssDNA) and double stranded DNA (dsDNA) fragments. This is because the observed charge of nucleic acids is proportional to the number of monomers (nucleotide or deoxynucleotide) plus the extra charge of the phosphate group at the 5' end of the fragment. This is always true when the molecule is located within a buffer solution in the 4 to 10 pH range. Below pH 4 the phosphate starts to neutralize and above pH 10 the bases start to protonate. For RNA and ssDNA this is given by $q = -(N + 1)e$, and for dsDNA it is given by $q = -2(N + 1)e$. Moreover, the electrostatic repulsive forces between each of these charged multiple segments make the strands migrate as open free-draining chains. They do not migrate as compact structures such as folded proteins do. Each monomer collides with the solvent while the solvent percolates through the random chain. Therefore, the total viscous friction coefficient is given by the number of monomers times the average friction coefficient of a monomer plus the friction coefficient of the neutral drag tag. The electrophoretic mobilities are given by:

$$\underbrace{\mu_e = \frac{-(N + 1)e}{Nf_s + f_{ut}}}_{\text{RNA or ssDNA}} \quad \text{and} \quad \underbrace{\mu_e = \frac{-2(N + 1)e}{Nf_d + f_{ut}}}_{\text{dsDNA}}, \tag{2.44}$$

where N represents the number of monomers, e represents the elementary charge, f_{ut} represents the uncharged drag tag friction coefficient, f_s represents the friction coefficient of a RNA or ssDNA monomer, and f_d represents the friction coefficient of a dsDNA monomer (pair of deoxinucleotides). Figure 2.22 shows the mobility of ssDNA as a function of N for many values of f_{ut}/f_s, the ratio of the uncharged drag tag friction coefficient to the friction coefficient of one monomer. Biotin modified oligonucleotides (with a biotin at the 3' (or 5') position) are very accessible. Adding streptavidin to the running buffer solution produces a drag tag (the biotin-streptavidin system), which exhibits a friction coefficient of $f_{ut}/f_s \simeq 13$.

2.15.4 Free-Solution Electrophoresis

The phenomenon discussed in Sections 2.3, 2.4, and 2.5 is called free-solution electrophoresis (FSE) (also known as zone electrophoresis). This is the base of the

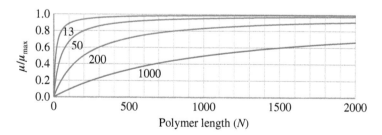

Figure 2.22 Variations of μ given as a function of the number of residues (N) of ssDNA, calculated for $f_{ut}/f_s = 13, 50, 200,$ and 1000. The biotin–streptavidin system has a typical value of $f_{ut}/f_s = 13$, i.e. it only results in a good separation of the short deoxinucleotides ($N < 150$).

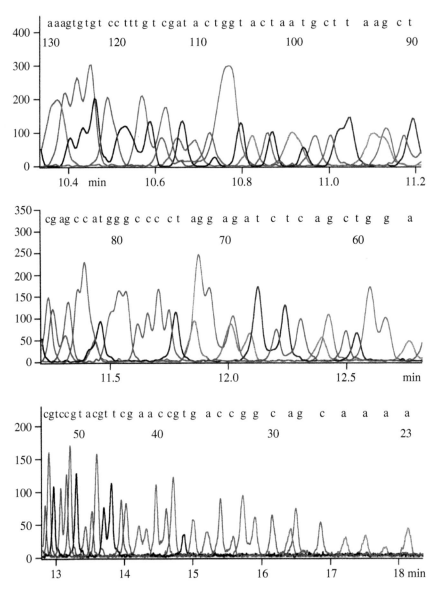

Figure 2.23 Example of a DNA sequencing run (electropherogram) using the ELFSE separation mode with the biotin–streptavidin system as the drag tag. Biotinilated primers and fluorescently labeled ddNTPs were used in the Sangers' chain-terminating reaction. Note the high resolution of the short deoxioligonucleotides and the lower resolution of the longer deoxioligonucleotides. This can be understood by examining equation 2.44 and Figure 2.22 (*Source*: Reproduced with permission of John Wiley & Sons).

simplest separation mode used to separate the components of interest (analytes) from samples in many fields (chemistry, biochemistry, genetics, life sciences, and many others). To achieve this, a narrow band of the sample can be applied to slabs of gel (the gel works merely as an anti-convective agent in this case); a small plug of the sample can be inserted into the inlet of capillaries or the sample plug can be inserted at the crossing of channels in microchips. A high voltage is immediately turned on and the electrophoretic velocity of the analytes is initially subjected to a very short transient regime (see Appendix C) in which the average electropheretic velocity goes from zero to the stationary state velocity. At the stationary state the electric force that pulls the ions is equal to the opposing viscous friction force. The resulting average net force acting on the ion or charged molecule will be zero and the resulting average electrophoretic velocity will stay constant over the remaining time, as long as the applied electric field remains constant. The velocity is given by $v = \mu_e E$, where μ_e is the electrophoretic mobility. The initial transient regime is ignored in most practical applications, as it is too short compared to the stationary state regime.

Free solution electrophoresis is the easiest to optimize in terms of separation efficiency, as pH plays the most important role. If non-derivatized analytes are being separated then a single plot of the mobilities (measured or from tables) against pH is sufficient to find the ideal pH. As an example, Figure 2.24 is a plot of the mobilities of three hypothetical amines.

In this case the EOF may be used to increase or reduce the run time in order to have all the analytes separated in the shortest time. Finally, a few questions still remain: which of these pH values are the best for the separation? If the number of analytes is large, how can someone decide the best pH value to separate all of them? There are several objective means of answering these. A simple one uses the productory of the square of the mobility differences. For the three analytes of

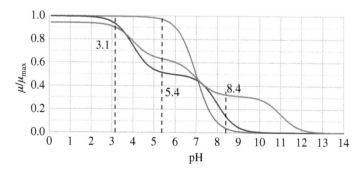

Figure 2.24 Mobilities of three hypothetical analytes divided by the maximum mobility of the three, plotted against pH in the 0 to 14 pH range. As shown in Figures 2.25 and 2.26 the optimal pH for the separations are 3.1, 5.4, or 8.4.

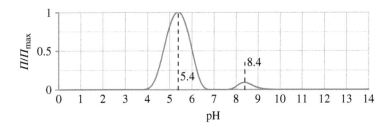

Figure 2.25 Productory of $(\mu_i - \mu_j)^2$ in the 0 to 14 pH range. The higher the productory the better. According to this criteria pH 5.4 is much better than pH 8.4 for this separation when not considering the effect of pH on the EOF, which can potentially be present.

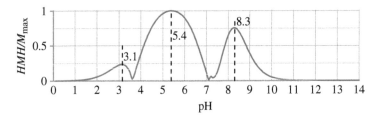

Figure 2.26 Harmonic mean of the modulus of all pairs of $|\mu_i - \mu_j|$ with $i \neq j$ in the 0 to 14 pH range. This shows similar values to Figure 2.25 in addition to a third window at around pH 3.1 that also allows, in principle, all of the components to be separated.

Figure 2.24 this productory is given by: $\Pi = (\mu_1 - \mu_2)^2(\mu_1 - \mu_3)^2(\mu_2 - \mu_3)^2/\Pi_{max}$, where Π_{max} is the maximum of the values found in the pH = 0 to 14 range. The resulting graph is shown in Figure 2.25. Alternatively, the harmonic mean (HM) among the modulus of all mobility differences $|\mu_i - \mu_j|$ also reveals the optimal pH for the separation. This reveals a third possible window to work which is located around pH = 3.1, as shown in Figure 2.26.

2.15.5 Isoelectric Focusing

The IEP is an important property of a molecule and is explained in Section 2.7. Isoelectric focusing is an electrophoresis separation mode that can be performed in any platform (capillary, microchip, or slab). It is characterized by the use of a pH gradient along the separation path, which focuses the bands at positions where pH = pH(I). This is usually made by using solutions of ampholytes and the positive voltage is applied to the reservoir with lower pH and the negative voltage to the reservoir with the higher pH. The distance between the resulting focused bands is proportional to the pH(I) differences among the analytes and pH gradient. By way of illustration, let us assume that the left band on the isoelectric focusing strip of Figure 2.27 is the amino acid glycine with pH(I) = 5.97. Note that the

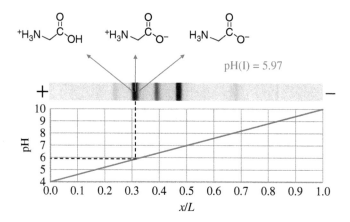

Figure 2.27 Isoelectric focusing of amino acids, peptides, or proteins using a pH gradient (in a microchannel, capillary or strip), within the pH 4 to 10 range, from $x = 0$ to L. If the indicated band is made of the amino acid glycine (pH(I) = 5.97) then three predominant ionic forms will be found along this focused band. The cation has a higher concentration than the anion on the left side of the band and the anion has a higher concentration than the cation on the right side of the band (author's unpublished results).

molecules of the band that are dispersed to the left (lower pH region) become positively charged and are driven back to the cross section points where pH = pH(I). Conversely, the molecules that diffuse to the right hand side, where pH > pH(I), start to experience a negative net charge (see Section 2.4). In this case a force acting in the opposite direction appears and the molecules are electrodriven back to any point of the cross sectional plane where pH = pH(I) = 5.97. Therefore, it does not matter how long the applied separation voltage is left on, the bands will always remain focused around these isoelectric points. The band widths are the result of two opposing phenomena called molecular diffusion and focusing forces, which compete between themselves. The first depends on the molecular diffusion coefficient (D) and the second depends on the spatial pH gradient (dpH/dx), the spatial voltage gradient (dV/dx), and the electrophoretic mobility. An interesting property of the technique is that the sample can be applied anywhere or can even be spread along the separation track (with a pH gradient). After a while the analytes will all migrate to the points were pH = pH(I) if the positive electrode is applied to the reservoir with the buffer with lower pH and the negative electrode to the buffer with higher pH. They all will stay focused around these points so long as the pH gradient and the voltage are maintained.

2.15.6 Isotachophoresis

The term isothacophoresis (ITP) comes from the joining of *iso* (the same), *thaco* (speed), and *phoresis*, which is a shortening of electrophoresis. This is exactly what happens in ITP when the system reaches equilibrium and all the components have been separated; the zones migrate at the same speed with spatial plateau-like concentration profiles (one plateau or zone for each component). To facilitate the description, a zero EOF will be considered. The zones reach the detection point in a decreasing mobility order (which serves as an identifying factor) and have different plateau widths (which are related to their concentrations). The way in which this technique is conduced creates huge electrical field changes along the separation medium. This is something odd, as the field is only approximately constant within each zone, and its intensity changes each time a front running component exits the separation media (capillary, microchannel, or slab) into the outlet reservoir.

To present this in an easy way, the analysis of anions and cations must be discussed separately. Two buffer solutions are used: the *terminating electrolyte* is used in the inlet reservoir and the *leading electrolyte* is used to fill the separation medium and the outlet reservoir (close to which the detectors are usually positioned). For the analysis of anions, the mobility of the anion of the leading electrolyte must be higher than the fastest moving anion of the sample, and the mobility of the anion of the terminating electrolyte must be slower than the slowest moving anion of the sample. When cations are analyzed then the mobility of the cation of the leading electrolyte must be larger than the fastest moving cation of the sample, and the mobility of the cation of the terminating electrolyte must be slower than the slowest moving cation of the sample.

Sample plugs with lengths that are one-tenth of the length of the separation medium can be injected. When anions are analyzed, the fastest anion of the sample will occupy the front zone, followed by the second, and so on. Strong field distortions will follow with strong non-linear couplings among the zones. During the run the electric field strength changes in both space and time along the separation medium. When the system reaches equilibrium a succession of analyte zones with well defined boundaries and no space in between move at the same speed. The electric field strength developed on each zone is given by: $E_1 \kappa_1 = E_2 \kappa_2 = E_3 \kappa_3 = \ldots = E_i \kappa_i$, i.e. the field strength is inversely proportional to the conductivity of the zone. This is because the ionic current must be constant through any cross section along the whole separation medium, because there are no sources and sinks of electric charges (ions).

When the molecules of any two neighbor zones start to mix due to molecular diffusion, a different field strength will build up in this third intermediate zone. The field strength of this zone will be based on its conductivity. As a result, the components will either speed up or slow down, thereby rejoining their original zone. This is a focusing effect that is specific to this technique and is different from that observed in isoelectric focusing (Section 2.15.5). Diluted components of the sample will be compressed because the solute concentration must satisfy this equation $E_i \kappa_i = E_j \kappa_j$ for any pair of anions i and j. This band compression phenomena promotes trace enrichment, which is perhaps one of the most interesting applications of ITP. Remember that the concentration is always proportional to the length rather than the height or area of the zones. This technique is also well suited for fraction collection, for instance of separated proteins. In this case it is common to add *spacer molecules* to the sample. These molecules must exhibit an intermediate mobility compared to the analytes of interest at the pH of the run, and must posses different detection characteristics. This facilitates fraction collection, as there will be an intermediate zone occupied by the spacer molecules.

2.15.7 Microemulsion Electrokinetic Chromatography

Microemulsions are suspensions of microdroplets, which are generally much larger than micelles (see Section 2.15.8), but are still smaller than 1 μm in diameter. These water-in-oil (W/O) or oil-in-water (O/W) suspensions are translucent, i.e. they have a low turbidity. They were discovered by Shulman and Hoar [80] and are easily produced using, for instance, water, octane (oil), SDS (surfactant), and butanol (co-surfactant). Firstly 0.8 g of octane is added to 90 mL of a 100 mM SDS aqueous solution. This gives a cloudy mixture under a stirrer, but it turns into a clear and stable suspension upon the addition of 6.5 g of 1-butanol. To this a few mL of 100 mM SDS solution is added, producing 100 g of the solution.

These microemulsions also work as pseudophases in a similar way to micelles in MEKC. They are mainly used in the capillary and microchip platforms. Oil droplets can be positively, neutrally or negatively charged, depending on the surfactant used. The more hydrophobic the analytes are, the more they will partition into the microemulsion droplets when the O/W microemulsions are used. Analytes are separated by two combined phenomena: the electrophoretic mobility and the mass distribution rate between the aqueous phase and the microemulsion pseudophase. There are recent reviews covering the application of MEEKC in many fields and it can also be combined with many sample stacking strategies to improve separation efficiencies and sensitivities [81,82].

2.15.8 Micellar Electrokinetic Chromatography

Micelles are nanoscopic supramolecular structures that are formed, for instance, when surfactants containing a hydrophilic head and a long hydrofobic tail are mixed in water above their critical micellar concentration. The surfactant's head may be neutral, anionic or cationic, and they can spontaneously aggregate in water to form spherical nanostructures with the hydrophilic heads in contact with the aqueous phase and the hydrophobic tails projected into the inner micellar phase. This forms a stable two phase system called colloidal suspension: an aqueous phase with electrolytes dissolved in it and a hydrophobic phase made by the micelles. This is known as normal-phase micelle (NP micelles used in NP MEKC or O/W MEKC), while the inverse is also possible: micelles with their head groups facing into the center and their tails extended outwards into the oil phase (RP micelles used in RP MEKC or W/O MEKC). Below the critical micelle concentration virtually no micelles are detected, and above it virtually all additional surfactant molecules form micelles. The sizes of the micelles are highly dependent on the structure of the surfactant, as well as the temperature and ionic strength of the aqueous phase [83,84].

These colloidal suspensions or mixtures are good at dissolving many classes of molecules, which exhibit poor solubility in pure water. Moreover, neutral analytes, which do not migrate in free solution electrophoresis, can migrate in these suspensions and are carried by the micellar mobile pseudophase if the micelles are made of cationic or anionic surfactants. This technique was named micellar electrokinettic chromatography (MEKC) [85,86] and is widely used, especially in the capillary and microchip platforms. The succession of events that led to the development of this very widely used EST was reported by Terabe [87] and there is a series of review articles covering the methodological and instrumental advances of MECK [88–90].

Figure 2.28 shows the separation of a mixture of proteinogenic amino acids using MEKC with SDS as the surfactant and borate buffer as the aqueous phase. In this case two distinct separation mechanisms are acting at the same time:

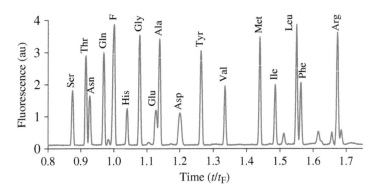

Figure 2.28 Separation of naphthalene-2,3-dicarboxaldehyde derivatized amino acids of a standard mixture using MEKC (author's unpublished results). The conditions used were as follows: 50 mM SDS in 25 mM borate buffer, with the pH adjusted to 9. A 67 cm long capillary, with an effective length of 65 cm, was used. It had a 50 μm ID and a 365 μm OD, and was cooled to 15 °C using a circulating liquid coolant. A voltage of 15 kV was applied and a current of 11.2 μA was observed. A LIF detector was operated at 405 nm and fluorescein was the internal standard (F) used. Migration time is expressed in relation to the fluorescein migration time (20.5 min). The concentrations of Ser, Asn, His, Glu, Asp, Val, Ile, and Phe were 4 μM and the concentrations of the other amino acids were 8 μM.

electrophoresis and the partitioning between the hydrophilic (aqueous) and the hydrophobic (micellar) pseudophase.

2.15.9 Sieving Electrophoresis

Sieving electrophoresis (SE) is normally conducted in gels or entangled polymer solutions. These matrices are used to separate, by length, polyanions and polycations. Examples of such charged polymers are RNA, ssDNA, dsDNA, charged oligosaccharides, and proteins. The proteins are first treated with a reducing agent to break the S–S bonds, and then with an anionic or cationic surfactant to denature them and to confer approximately one charge per residue (one surfactant molecule interacts with each residue). This converts the protein into a polyion-like chain. While electrophoresis still remains the predominant migration phenomena in these separations, longer molecules have a higher collision rate with obstacles (gel or polymer), which decreases their relative mobilities and produces separations based on analyte length.

It is worth to remember that "sieves" are devices with meshes or pores through which finer particles of a mixture are separated from coarser ones. They have a clear cut-off defined by the mesh or pore size. This is not the predominant separation mechanism in sieving electrophoresis, despite the presence of pores in gels. The separation matrices used in SE can be 2D array of microfabricated posts, membranes, cross linked or branched polymers (gels), concentrated polymer solutions (highly entangled), semi-dilute polymer solutions (entangled), and also dilute polymer solutions (non-entangled). The latter do not have pores, but still permit a relatively good separation. Therefore, the important separation mechanisms used in SE are the distinct collision rates of the analytes (polyions) against the neutral obstacles (microfabricated posts, gel net, or neutral polymers) and the distinct disentangling times of the polyions from these obstacles. Despite this, the word "sieving" electrophoresis became of common use to name this separation mode. This happened probably due to the lack of a shorter, more accurate alternative name for it.

When any non-interacting neutral gel presents an average pore size that is much larger than the radius of gyration of the moving entities, then equations 2.2 are still valid. In these cases the gel works merely as an anti-convective agent. Therefore, the separation mode will be free solution electrophoresis and not sieving electrophoresis [91]. The separation will only be sieving electrophoresis if the average radius of gyration of the migrating molecules are of the order of, or larger than, the average pore size of the gel. The relative pore sizes of the gel, compared to the radius of gyration of the analytes, leads to a few migration regimes [92–94], which, at constant field strength, are called: (1) Ogston sieving (the random coil DNA is sieved and mobility decreases exponentially with length), (2) entropic trapping (the random coil jumps between pores), (3) near-equilibrium reptation (the random coil migrates head first through the gel), (4) reptation trapping, (5) oriented trapping, (6) geometration, and (7) complete trapping.

Gels are commonly used in the slab platform for a great variety of applications, ranging from DNA sizing, verifying RNA integrity, protein sizing, 2D electrophoresis in proteomics (first isoelectric focusing is performed in one dimension and then gel electrophoresis is performed in the second dimension to size proteins), separation of very long DNA fragments (pulsed field electrophoresis in agarose gels), and many others. After the run the samples must be stained to make the bands visible. DNA is stained with fluorophores such as ethidium bromide, which fluoresce under ultraviolet light when intercalated into the major grove of dsDNA. Alternatively, proteins are visualized using silver stain, coomassie brilliant blue, and others.

The large demand for DNA sequencing and the difficulty of replacing gels in microchannels and capillaries has led to the development of flowable polymer solutions. These solutions have a high viscosity and require special accessories in the instruments for their use, such as: syringes with servo-motors or high pressure nitrogen. Moreover, sensitive fluorescence detectors are required and dedicated software (basecallers). Read lengths over one thousand have been achieved in less than one hour using linear poly(acrylamide) as the sieving matrix [95,96] (see Figure 2.29). A "four color" chemistry [97] was also developed so that all four bases can be read in a single capillary. Prototypes of instruments capable of working with this new method were simultaneously developed [98]. Using such polymer solutions, medium sized dsDNA (50–1000 bp) is also routinely separated, but even the separation of much larger molecules (up to 50 kbp) is possible.

2.15.10 Suitable Separation Modes for Each Class of Analytes

Within the literature it is possible to see that the separation mode used is often not optimal for the particular application. There are many explanations for this, but the most common reason is that authors prefer to use the separation mode that they are most familiar with. Another contributing factor may be the cost of the instruments. Nevertheless, it is useful to compare all of the separation modes and determine their suitability for each class of analytes and sample type. This is summarized in Table 2.4; however in some cases it is difficult to predict (when starting from first principles) the separation mode that will give the best results in the shortest run time. Therefore, trial and error is still sometimes used to determine which separation mode is the best.

There are many important exceptions that do not fit well in Table 2.4. Here are some examples: a small molecule with LogP ≪ 0 and an IEP can be separated by IEF; a charged molecule with LogP ≫ 0 (for instance, quaternary amines)

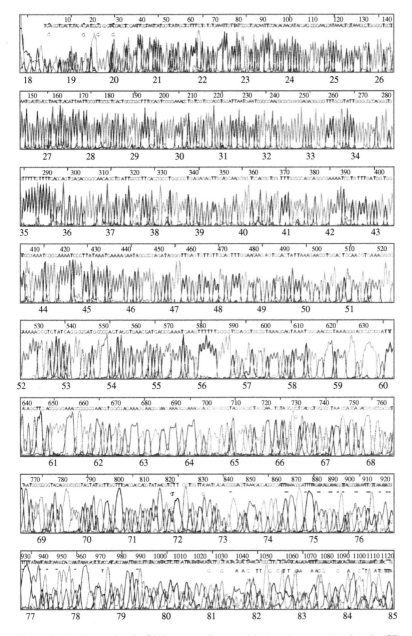

Figure 2.29 Example of a DNA sequencing run (electropherogram) using the SE separation mode in a polymer solution (linear polyacrilamide). Oligonucleotides were produced by Sanger's termination reaction using AmpliTaq-FS cycle sequencing and labeled primers on ssM13mp18. Conditions: electric field 150 V cm^{-1}, running buffer 50 mM Tris/50 mM TAPS/2 mM EDTA/7 M urea. Temperature 50 °C, 2% linear polyacrilamide, and PVA coated capillary with 30 cm effective length and 45 cm total length. (*Source*: Reproduced with permission of American Chemical Society).

Table 2.4 The most appropriate separation modes for each class of analytes when considering also their LogP values.

Analyte class	LogP	Separation modes
Small ions	—	FSE, ITP
Small molecules	$\ll 0$	FSE, O/W MEEKC or MEKC, ITP
Small molecules	$\simeq 0$	O/W or W/O MEEKC or MEKC
Small molecules	$\gg 0$	W/O MEEKC, W/O MEKC
Peptides	—	FSE, IEF, O/W MEKC or MEEKC
Proteins	—	FSE, IEF
Oligosaccharides	—	ELFSE, FSE
Oligonuceotides	—	SE, ELFSE

can be separated by FSE using organic solvents as the separation medium instead of aqueous solutions; reducing mono, di-, and oligosaccharides can be separated by FSE after derivatization with 8-aminopyrene-1,3,6-trisulfonic acid; the enantiomeric separation of some molecules require the use of a combination of MEKC (SDS) and AE (cyclodextrins or other chiral selector additives); to cite a few. Finally, it is important to mention that the nature of the starting sample (sometimes called matrix) is also very important while deciding on the best separation modes. Moreover, most of the samples require important preparation steps before they are ready for injection and analysis; however this is out of the scope of this book.

References

1 Delgado, A.V., González-Caballero, F., Hunter, R.J. et al. (2007). *Journal of Colloidal and Interface Science* 309: 194–224.
2 Vesterberg, O. (1989). *Journal of Chromatography* 480: 3–19.
3 Vesterberg, O. (1993). *Electrophoresis* 14: 1243–1249.
4 Jackson, J.D. (1999). *Classical Electrodynamics*, 3e. New York: Wiley.
5 Shera, E.B., Seitzinger, N.K., Davis, L.M. et al. (1990). *Chemical Physics Letters* 174: 553–557.
6 Zusková, I., Novotná, A., Vcelaková, K., and Gaš, B. (2006). *Journal of Chromatography B* 841: 129–134.
7 Koval, D., Kašička, V., and Zusková, I. (2005). *Electrophoresis* 26: 3221–3231.
8 Jaroš, M., Včeláková, I., and Gaš, B. (2002). *Electrophoresis* 23: 2667–2677.
9 Li, D., Fu, S., and Lucy, C.A. (1999). *Analytical Chemistry* 71: 687–699.

10 Balashov, V.N., Hnedkovsky, L., and Wood, R.H. (2017). *Journal of Molecular Liquids* 239: 31–44.

11 Saar, K.L., Müller, T., Charmet, J. et al. (2018). *Analytical Chemistry* 90: 8998–9005.

12 Malá, Z. and Gebauer, P. (2019). *Electrophoresis* 40: 55–64.

13 Malá, Z., Gebauer, P., and Boček, P. (2017). *Electrophoresis* 38: 9–19.

14 Malá, Z., Gebauer, P., and Boček, P. (2015). *Electrophoresis* 36: 2–14.

15 Malá, Z., Gebauer, P., and Boček, P. (2013). *Electrophoresis* 34: 34–18.

16 Gebauer, P., Malá, Z., and Boček, P. (2011). *Electrophoresis* 32: 83–89.

17 Righetti, P.G. (2004). *Journal of Chromatography A* 1037: 491–499.

18 Herzog, C., Beckert, E., and Nagl, S. (2014). *Analytical Chemistry* 86: 9533–9539.

19 Łapińska, U., Saar, K.L., Yates, E.V. et al. (2017). *Physical Chemistry Chemical Physics* 19: 23060–23067.

20 Reijenga, J., van Hoof, A., van Loon, A., and Teunissen, B. (2013). *Analytical Chemistry Insights* 8: 53–71.

21 Mai, T.D. and Hauser, P.C. (2011). *Talanta* 84: 1228–1233.

22 Qian, C., Wang, S., Fu, H. et al. (2018). *Electrophoresis* 39: 1786–1793.

23 Jarvas, G., Szigeti, M., and Guttman, A. (2018). *Journal of Separation Science* 41: 2473–2478.

24 Lambert, W.J. and Middleton, D.L. (1990). *Analytical Chemistry* 62: 1585–1587.

25 Lucy, C.A., MacDonald, A.M., and Gulcev, M.D. (2008). *Journal of Chromatography A* 1184: 81–105.

26 Hjertén, S. (1985). *Journal of Chromatography* 347: 191–198.

27 Hajba, L. and Guttman, A. (2017). *TrAC Trends in Analytical Chemistry* 90: 38–44.

28 Liao, J.L., Abramson, J., and Hjertén, S. (1995). *Journal of Capillary Electrophoresis* 2: 191–196.

29 Gilges, M., Kleemlss, M.H., and Schomburg, G. (1994). *Analytical Chemistry* 66: 2038–2046.

30 Belder, D., Deege, A., Husmann, H. et al. (2001). *Electrophoresis* 22: 3813–3818.

31 Suratmann, A. and Wätzig, H. (2007). *Electrophoresis* 28: 2324–2328.

32 Stevens, T.S. and Cortes, H.J. (1983). *Analytical Chemistry* 55: 1365–1370.

33 Lauer, H.H. and McManigill, D. (1986). *Analytical Chemistry* 58: 166–170.

34 Huang, X.H., Gordon, M.J., and Zare, R.N. (1988). *Analytical Chemistry* 60: 1837–1838.

35 Vandegoor, A., Wanders, B.J., and Everaerts, F.M. (1989). *Journal of Chromatography* 470: 95–104.

36 Altria, K.D. and Simpson, C.F. (1987). *Chromatographia* 24: 527–532.

37 Wang, W., Zhao, L., Jiang, L.P. et al. (2006). *Electrophoresis* 27: 5132–5137.

38 Wang, W., Zhou, F., Zhao, L. et al. (2007). *Journal of Chromatography A* 1170: 1–8.

39 Zhang, W., He, M., Yuan, T., and Xu, W. (2017). *Electrophoresis* 38: 3130–3135.

40 Sutherland, W. (1905). *Philosophical Magazine* 9: 781–785.

41 Einstein, A. (1905). *Annalen der Physik* 322: 549–560.

42 von Smoluchowski, M. (1906). *Annalen der Physik* 326: 756–780.

43 Bockris, J.O'M. and Reddy, A.K.N. (1998). *Modern Electrochemistry*, vol. 1, 2e, 505–526. Plenum Press: New York.

44 Khaledi, M.G. (ed.) (1998). *High-Performance Capillary Electrophoresis: Theory, Techniques and Applications*, 1047. New York: Wiley.

45 Stellwagen, E. and Stellwagen, N.C. (2002). *Electrophoresis* 23: 2794–2803.

46 Ghosal, S. (2004). *Electrophoresis* 25: 214–228.

47 Poppe, H. (1992). *Analytical Chemistry* 64: 1908–1919.

48 Kašička, V., Prusík, Z., Gaš, B., and Štědrý, M. (1995). *Electrophoresis* 16: 2034–2038.

49 Sonke, J.E., Furbish, D.J., and Salters, V.J.M. (2003). *Journal of Chromatography A* 1015: 205–218.

50 Cao, J., Hong, F.J., and Cheng, P. (2007). *International Communications in Heat and Mass Transfer* 34: 1048–1055.

51 Gaš, B., Štědrý, M., Rizzi, A., and Kenndler, E. (1995). *Electrophoresis* 16: 958–967.

52 Štědrý, M., Gaš, B., and Kenndler, E. (1995). *Electrophoresis* 16: 2027–2033.

53 Terabe, S., Shibata, O., and Isemura, T. (1991). *Journal of High Resolution Chromatography* 14: 52–55.

54 Malá, Z., Křivánková, L., Gebauer, P., and Boček, P. (2007). *Electrophoresis* 28: 243–253.

55 Malá, Z., Šlampová, A., Gebauer, P., and Boček, P. (2009). *Electrophoresis* 30: 215–229.

56 Malá, Z., Gebauer, P., and Boček, P. (2011). *Electrophoresis* 32: 116–126.

57 Šlampová, A., Malá, Z., Pantčková, P. et al. (2013). *Electrophoresis* 34: 3–18.

58 Malá, Z., Šlampová, A., Křivánková, L. et al. (2015). *Electrophoresis* 36: 15–35.

59 Šlampová, A., Malá, Z., Gebauer, P., and Boček, P. (2017). *Electrophoresis* 38: 20–32.

60 Šlampová, A., Malá, Z., and Gebauer, P. (2019). *Electrophoresis* 40: 40–54.

61 Mandaji, M., Rübensam, G., Hoff, R.B. et al. (2009). *Electrophoresis* 30: 1501–1509.

62 Mandaji, M., Rübensam, G., Hoff, R.B. et al. (2009). *Electrophoresis* 30: 1510–1515.

63 Yu, F., Zhao, Q., Zhang, D. et al. (2019). *Analytical Chemistry* 91: 372–387.

64 Dubský, P., Dvořák, M., and Ansorge, M. (2016). *Analytical and Bioanalytical Chemistry* 408: 8623–8641.

65 Pretorius, V., Hopkins, B.J., and Schieke, J.D. (1974). *Journal of Chromatography* 99: 23–30.

66 Jorgenson, J.W. and Lukacs, K.D. (1981). *Journal of Chromatography* 218: 209–216.

67 Vessman, J., Stefan, R.I., van Staden, J.F. et al. (2001). *Pure and Applied Chemistry* 73: 1381–1386.

68 D'Orazio, G., Asensio-Ramos, M., Fanali, C. et al. (2016). *TrAC Trends in Analytical Chemistry* 82: 250–267.

69 Hu, W., Hong, T., Gao, X., and Ji, Y. (2014). *TrAC Trends in Analytical Chemistry* 61: 29–39.

70 Tarongoy, F.M. Jr., Haddad, P.R., and Quirino, J.P. (2018). *Electrophoresis* 39: 34–52.

71 Mayer, P., Slater, G.W., and Drouin, G. (1994). *Analytical Chemistry* 66: 1777–1780.

72 Sudor, J. and Novotny, M.V. (1995). *Analytical Chemistry* 67: 4205–4209.

73 Sudor, J. and Novotny, M.V. (1997). *Analytical Chemistry* 69: 3199–3204.

74 Ren, H., Karger, A.E., Oaks, F. et al. (1999). *Electrophoresis* 20: 2501–2509.

75 Vreeland, W.N. and Barron, A.E. (2000). *American Chemical Society, Polymer Preprints, Division of Polymer Chemistry* 41: 1018–1019.

76 Won, J.-I., Meagher, R.J., and Barron, A.E. (2005). *Electrophoresis* 26: 2138–2148.

77 Meagher, R.J., Won, J.-I., Coyne, J.A. et al. (2008). *Analytical Chemistry* 80: 2842–2848.

78 Haynes, R.D., Meagher, R.J., Won, J.-I. et al. (2005). *Bioconjugate Chemistry* 16: 929–938.

79 Istivan, S.B., Bishop, D.K., Jones, A.L. et al. (2015). *Analytical Chemistry* 87: 11433–11440.

80 Hoar, T.P. and Shulman, J.H. (1943). *Nature* 152L: 102–103.

81 Ryan, R., Altria, K., Mcevoy, E. et al. (2013). *Electrophoresis* 34: 159–177.

82 Yang, H., Ding, Y., Cao, J., and Li, P. (2013). *Electrophoresis* 34: 1273–1294.

83 Myers, D. (ed.) (2006). *Surfactant Science and Technology*, 3e. Hoboken, NJ: Wiley.

84 Cosgrove, T. (2010). *Colloidal Science – Principle, Methods and Applications.* Hoboken, NJ: Wiley.

85 Terabe, S., Otsuka, K., Ichikawa, K. et al. (1984). *Analytical Chemistry* 56: 111–113.

86 Terabe, S., Otsuka, K., and Ando, T. (1985). *Analytical Chemistry* 57: 834–841.

87 Terabe, S. (2010). *Procedia Chemistry* 2: 2–8.

88 Silva, M. (2013). *Electrophoresis* 34: 141–158.

89 Silva, M. (2011). *Electrophoresis* 32: 149–165.

90 Silva, M. (2009). *Electrophoresis* 30: 50–64.

91 Shimizu, T. and Kenndler, E. (1999). *Electrophoresis* 20: 3364–3372.

92 Slater, G.W., Mayer, P., and Drouin, G. (1998). *Methods in Enzymology* (ed. B.L. Karger). San Diego, CA: Academic Press.

93 Viovy, J.-L. (2000). *Reviews of Modern Physics* 72: 813–872.

94 Slater, G.W. and Noolandi, J. (1986). *Biopolymers* 25: 431–454.

95 Carrilho, M., Ruiz-Martinez, M.C., Berka, J. et al. (1996). *Analytical Chemistry* 68: 3305–3313.

96 Salas-Solano, O., Carrilho, O., Kotler, L. et al. (1998). *Analytical Chemistry* 70: 3996–4003.

97 Lee, L.G., Spurgeon, S.L., Heiner, C.R. et al. (1997). *Nucleic Acids Research* 25: 2816–2822.

98 Kheterpal, I., Scherer, J.R., Clark, S.M. et al. (1996). *Electrophoresis* 17: 1852–1859.

3

Open Layout

3.1 Introduction

In this chapter the platforms with an open (common) layout will be reviewed, including open capillary electrophoresis, open microchip electrophoresis, and open slab electrophoresis. More details of this systematic nomenclature are given in Appendix A. The performance indicators, which are identical for these three platforms, are calculated in a way that allows them to be compared with the performances of the toroidal layouts that will be described in Chapter 4.

3.2 Capillary Electrophoresis

In 1967 Hjertén [1] studied free-solution electrophoresis using quartz microtubes (with an ID of 300 µm). He produced a long and rich paper detailing the instrumental setup used, and presented an extended theoretical treatment of what was, at that time, called "free zone electrophoresis". Further progress was reported in a second paper three years later [2]. In 1974 Virtanen took another important step towards what is currently known as capillary electrophoresis (CE) by performing, for the first time, free-solution electrophoresis in 200 µm ID borosilicate tubes, using a potentiometric detector to register the separated alkali cations [3]. This was the subject of his PhD thesis at the Helsinki University of Technology. In the same year Pretorius et al. [4] ran benzene and acetone in 1 mm ID quartz tubes packed with microparticulate silica, applying 5 kV to columns ranging from 5 cm to 1 m in length. They demonstrated that voltage driven (electrochromatography) separations produced bands with half the plate height of those produced by pressure driven (liquid chromatography) when the same columns and packing materials are used. Five years later Mikkers and Everaerts performed free-solution electrophoresis in 200 µm ID poly(tetrafluoroethylene) microtubes [5]. They studied the adverse effects of overloading samples containing solutes with a mobility

Open and Toroidal Electrophoresis: Ultra-High Separation Efficiencies in Capillaries, Microchips, and Slabs, First Edition. Tarso B. Ledur Kist.
© 2020 John Wiley & Sons Ltd. Published 2020 by John Wiley & Sons Ltd.

that is higher or lower than that of the carrier's (BGEs) constituents [6]. In the 1980s Jorgenson and Lukacs succeeded in performing free-solution electrophoresis in borosilicate glass microtubes with an ID of only 75 μm, using an electrostatic potential difference of 30 kV and laser induced fluorescence detection. They demonstrated separation efficiencies of a few hundred thousand plates for a series of applications [7–10].

The microtubes used by Jorgenson and Lukacs were made of borosilicate glass. However, the fabrication of fused silica optical fibers by pulling a preform rod in "draw towers" was already a mature technology at that time. The production of fused-silica capillaries is a small step away from the optical fiber fabrication process. This was achieved by Bente et al. at Hewlett-Packard [11] in the same year as Jorgenson and Lukacs's first publication [7]. The original intended use of fused silica microtubes was as columns for gas chromatography, where long columns are sometimes used. For instance, a fused-silica tube preform with a 2 mm ID, 15 mm OD, and a length of 1 m can produce kilometers of microtube with an ID of 50 μm and an OD of 375 μm when heated close to its melting point inside a furnace located at the top of a drawing tower. The ID of the microtube is held constant during the pulling process by automatically adjusting the pulling speed. Before the microtubes are wound onto a spool they receive one or more external layers of polymer coating (e.g., poly(imide)), totaling 10 μm to 20 μm in thickness, for mechanical protection (suppression of microbends). The end product is a *flexible fused-silica microtube*, popularly known as a capillary, which is shown in Figure 3.1. A meter of this capillary is very cheap compared to a HPLC stainless steel packed column.

Many lasers also became commercially available in the 1980s, which facilitated the detection of minute quantities of solutes in the picoliter volumes of a detection cell. As an example, a 10 μm long detection cell that is used on a 50 μm ID capillary has a volume of only 20×10^{-12} L (20 pL), which is one-thousandth of

Figure 3.1 Schematic illustration of the shape of the capillary platform, used in the open layout called *open capillary electrophoresis* or simply *capillary electrophoresis*. The word "capillary" is a shortening of *flexible fused-silica microtube*. The ID used range from 20 to 100 μm, with 50 and 75 μm being the most used. The added external polymer coating, which is usually 10 to 20 μm thick, makes the capillary flexible and therefore prevents breakages.

a nanoliter. Laser induced fluorescence detection in such small volumes is possible for solutes with native fluorescence or, for molecules which are not naturally fluorescent, either fluorogenic reactions or labeling reactions to add a fluorescent tag to non-fluorescent molecules may be used. Last but not least, data acquisition boards and personal computers also became available during this time period. These technologies are important because data acquisition boards convert the analogic output of detectors into a digital format (using A/D converters) and personal computers speed up data acquisition (electropherograms) and processing (integration).

The work of Jorgenson and Lukacs can be considered to have been published at the right time as the previously cited technologies that became available during this period of time facilitated the replication of their EST by many others. Consequently, CE developments boomed [12] and applications were developed across many fields including pharmaceutical analysis [13,14], clinical and forensic analysis [15], food analysis [16,17], mono-, di-, oligo-, and polysaccharides [18,19], metabolomics [20,21], amino acids [22–24], peptides [25,26], proteins [27–29], and nucleic acid analyses (genotyping, methylation analysis, sequencing, and aptamer selection) [30]. Perhaps one of the most impressive applications was the first sequencing of the whole human genome, which was achieved by two independent groups [31,32]. CE became a well established and successful technique [12] and still with great potential in many fields [33].

Fused silica capillaries are known for their following beneficial properties:

1. Very strong (tensile strength).
2. Very flexible (when coated).
3. Chemically inert interior surface.
4. Mirror smooth interior surface.
5. Transparent from the deep UV to near IR (coating must be removed).
6. Easily cleaved and cut.
7. Very high breakdown voltage.
8. Very high electric resistivity.
9. Supports very high temperatures.
10. Thermal conductivity higher than almost all themoplastics.
11. Constant ID and OD over long lengths.
12. Internal surface modification is possible.
13. Fusion splicing to form joints is possible.

Figure 3.2 shows a capillary installed on an instrument with optics, reservoirs, and electrodes. It is common to operate instruments with many capillaries in parallel; some instruments with 8, 16, or even 96 capillaries can be used simultaneously. Dedicated programs automatically load the trays with samples, inject the samples, run the samples, and process the data.

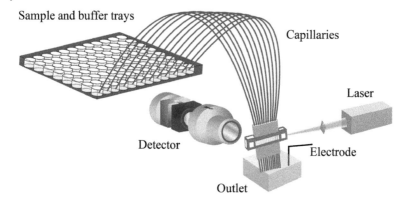

Sample and buffer trays

Capillaries

Laser

Detector

Electrode

Outlet

Figure 3.2 Schematic illustration of an array of capillaries mounted on a CE instrument showing the electrodes, reservoirs, and detector.

3.3 Microchip Electrophoresis

Separations performed along microchannels and electric potential gradients began more than thirty years ago [34–36]. These platforms were initially called micro-total analysis systems (μTAS), later they were called lab-on-a-chip (LOC), and finally received the name *microchip electrophoresis* (ME). In this platform the microchannels are fabricated on top of flat and smooth dielectric materials using photolithography, soft-lithography, toner-mediated lithography [37], or hot-embossing (also called microthermoforming) [38,39]. When the microchannels are ready a cover slide with holes, which will be used as buffer and sample reservoirs, is laid over the channels and permanently fused in place, as shown in Figure 3.3. More recently 3D printing has been successfully used to produce

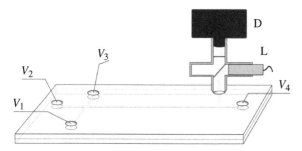

Figure 3.3 Schematic illustration of the setup on a typical microchip platform showing the microchannels and the wells where the buffer and samples are placed. This is the open layout version of the microchip platform, also called *open microchip electrophoresis* or simply *microchip electrophoresis* (ME). Dozens of separation modes, defined by the solutions and additives used to fill the microchannels and reservoirs, can be performed on this platform. The electrodes (V_1 to V_4) and a laser induced fluorescence detector, with a detector (D) and a laser (L), are also shown. Many other detection systems are compatible with this platform.

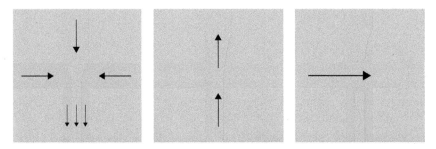

Figure 3.4 Diagrammatic representation of the microfluidic operation procedure used to inject the samples for ME. First the sample is electrodriven from reservoir 3 to reservoir 1 (see Figure 3.3) with the addition of a small flux of buffer from reservoirs 1 and 4. It is then driven back from reservoir 1 to reservoir 3 for a short time before being set to run in the direction of reservoir 4.

microchannels and their reservoir [40,41], adding flexibility to the design of microchannel based separation processes (microchips) [42,43].

Figure 3.3 shows the setup and parts of ME and Figure 3.4 shows the sample injection procedure [44]. This platform is known for its poor heat dissipation efficiency due to the materials used in their fabrication. The short track lengths (where the runs are performed) have also been cited as an disadvantage as they limit the achievable resolutions and peak capacities. However, in compensation, they have an enormous flexibility, are small (take small spaces in the laboratories), they consume small amounts of solvents and reagents, and many pre-run and post-run operations can be performed on-board, making them a great choice for an uncountable number of applications in many fields. The number of detectors that can be installed is unparalleled. Examples of detection systems that can be used are: capacitively coupled contactless conductivity [45], conductimety and potentiometry [3,46], amperometry[47], voltametry [48] (for a review of the main electrochemical detectors see the publication of Vandaveer et al.[49]), chemiluminescence [50], fluorescence [7,51,52], scanning laser induced fluorescence [53], time-resolved fluorescence [54], conductivity-based photothermal absorbance [55], electrochemiluminescence [56], mass spectrometry [57–59], and absorbance [60–64], to mention a few.

Microchip electrophoresis is expected to boom in many niches of applications during the first decades of the 21st century. This comes from many operations that can be performed on this platform, such as mixing nanoliters in microseconds [65], chemical and biochemical microreactions [66], and cell lysis and PCR reactions [67–69]. This increases the number of applications in many areas [36,70], namely DNA sequencing [71], protein sizing[72], proteomics and pepdidomics [73–75], IEF of proteins [76], peptides[77], biomedical analyses [78], microdialysis[79], ion analyses [80,81], amino acids [82–87], pharmaceutical analyses [88], food analyses [89,90], metabolites [91], and saccharides[92], to mention a few.

3.4 Slab Electrophoresis

In the slab platform (Figure 3.5) electrodriven separations occur within a slab of material, used as a support, which is laid in an electrophoresis chamber as shown in Figure 3.6. The slabs used in *open slab electrophoresis*, simply called *slab electrophoresis*, are typically a few millimeters thick (2 to 10 mm) and dozens of centimeters wide and long. The most common support materials are gels, nylon membranes, and cellulose acetate membranes. Their primary function is to work as anti-convective agents, i.e. they prevent band dispersion due to the convection of mass. In some applications these materials are designed to actively participate

Figure 3.5 The slab platform in the open layout, or *open slab electrophoresis*. Schematic illustration of the shape of a gel slab, showing the wells on top where samples are inserted and the migrating bands (the left array may be the reference bands). The slabs are typically 1 to 10 mm thick, 10 to 30 cm wide, and 10 to 50 cm long. They may be operated while lying horizontally in a tray or standing vertically. In the later case one buffer tank with electrodes is placed at the top and a second one is placed at the bottom.

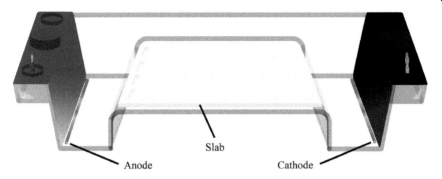

Figure 3.6 Perspective view of a vertical cut of an electrophoresis chamber used to conduct voltage driven separations within slabs. The electrodes and the two buffer reservoirs are shown. A cover is usually used to protect the users from electric shocks.

in the separation. One example is the sieving electrophoresis separation mode, where slabs of gels (e.g. poly(acrylamide)) are used to separate nucleic acids and proteins by length. If the DNA fragments are too long an agarose gel is used because of its larger pore sizes. This separation mode is popularly called "gel electrophoresis", while a more precise nomenclature would define it as *open slab sieving electrophoresis* (OSSE). When used to separate proteins it is known as "poly(acrylamide) gel electrophoresis with dodecyl sulfate" (SDS-PAGE). In this case SDS denatures the proteins and gives them a negative charge (giving an average elementary charge of 1.4 per residue), making them migrate as polyanions. The gel plays a fundamental role as it works as an obstacle, slowing the movement of longer fragments (as they have a higher collision rate) and facilitating the estimation of their length (molecular weight) (further explained in Section 2.15.9).

Note that most of the steps of slab electrophoresis separations are performed manually. These steps are:

(1) The melted viscous solution (normal agarose gel melts at 60 °C and "low melt" agarose melts at 40 °C) is poured into a mold where a comb is used to produce wells where samples will be injected. Alternatively, the gel can be polymerized in a casting mold. For instance, acrylamide is co-polymerized with N,N'-methylene-bisacrylamide. This is a radical co-polymerization in which the N,N'-methylene-bisacrylamide monomer creates branches during the polymerization, producing a solution that has a consistent structure; this is a typical property of gels.

(2) When the mixture is cold (or polymerized) and firm, the slab is laid into the electrophoresis chamber and the buffer solution is added to the reservoirs until it overflows and covers the slab.

(3) The samples are loaded into the wells using a micropipette and the voltage is immediately turned on to begin the run.

(4) When the run is finished the voltage is turned off, the gel is stained to facilitate band detection, and the bands either photographed (fluorescent stains are photographed in a transilluminator where a UV lamp is used for excitation) or read by a scanning device.

It is important to note that many separation modes can easily be performed in this slab platform, in both the open and toroidal layouts (See Chapter 4). This is accomplished by using gels with large pore sizes (e.g. diluted agarose gels). If the radius of gyration of the analytes is much smaller than the average pore size of the gel then it is no longer sieving electrophoresis, since the gel has a negligible effect on the separation mechanism [93]. In this case many separation modes can be performed, for instance affinity electrophoresis, end-labeled free solution electrophoresis, free solution electrophoresis, isoelectric focusing, micellar electrokinetic chromatography, and microemulsion electrokinetic chromatography, among others. However, renewing or exchanging the background electrolyte (BGE) between runs may be difficult in some of these separation modes. Their separation efficiencies are generally lower than those obtained with the capillary and microchip platforms because the applied electric field strength is limited to a few dozen volts per centimeter, depending on the thickness of the slab used and the heat removal efficiency of the system (slab+chamber).

The main advantages of SE are its convenience, simplicity, and that the separated bands can be collected for further treatment and analysis. The bands are collected using a cutting device or a hollow rectangular "hole punch". Alternatively, protein and RNA/DNA bands can be transferred from the gel onto a membrane (a procedure called blotting) and further analyzed (e.g., by antibody recognition for proteins or hybridization for RNA/DNA). Even further, direct transfer (or blotting) procedures have been developed, combining separation and transfer into a single step and thus eliminating a number of manual procedures. Out of all of the ESTs only free-flow electrophoresis rivals it in terms of the quantity of sample that can be collected following the run. Alternatively the presence of metaloproteins and phosphorous containing proteins can be analysed directly on the slab using laser-induced breakdown spectroscopy and other hand-held instruments [94].

Slab electrophoresis is among the oldest ESTs and is still being improved [95] and with important variants such as two-dimensional gel electrophoresis (an IEF in the first direction followed by a SDS poly(acrylamide) gel electrophoresis of the proteins in the orthogonal direction) [96,97], fluorescent two-dimensional difference gel electrophoresis (two or more samples are labeled with different fluorophores and then run together) [98,99], poly(acrylamide) gradient electrophoresis (the gel concentration grows along the separation direction) [100], pulsed-field electrophoresis [101], and single cell gel electrophoresis assay (comet assay) [102,103].

3.5 Performance Indicators for Open Layouts

Quantitative parameters allow the intrinsic capability of the different separation methods to be compared in an objective manner. One such parameter is the *number of theoretical plates* (N), a dimensionless parameter that is also known as separation efficiency. It does not take into account the time needed to deliver a given number of plates. Therefore the *number of theoretical plates per unit time squared* (plate double-rate) is an even more interesting parameter, as shown in Chapter 5. *Plate height* (H) is another way to express separation efficiency as it is inversely related to plate number. These terms (plate number and plate height) have been borrowed from the field of distillation towers, which is also a separation technique, and are now applied to ESTs, chromatography, and sedimentation techniques [104,105]. The *resolution* (R) of two neighboring bands (along the separation track in capillaries, microchannels, or slabs) or peaks (along the electropherogram) is another interesting, dimensionless parameter that indicates how well two neighbors are separated or resolved. Finally, *band capacity* (B) and *peak capacity* (P) indicates the maximum number of analytes that can be separated per run. This is made by counting the number of bands along the separation path (with the aid of a camera or scanner) or the number of peaks in the electropherogram (produced by a detector positioned and immobile at the end of the separation path). B represents the number of bands detected along the separation path at a given time (at time t_m with the aid of a camera or scanner), while P represents the number of peaks that appear in an electropherogram (produced by a detector positioned and immobile at the end of the separation path at $x = l$). The calculation of P is more complicated than B and they must be separated into two distinct sections.

Note that the spatial profile of the bands along the separation tracks are easily measured using spatial scanners (in capillaries or microchannels) [106], which detect the spatial profile of the bands at a given time, or by photographic cameras within transilluminators (in slabs).

Regardless of which platform or separation mode is used, separation performances depend either on the distance run by the center of mass of the bands and on the variances of the bands or on the migration time of the centers of mass of the peaks shown in the electropherogram and on the variances of the peaks that are also shown in the electropherogram. The only exception to this is isotachophoresis, as within this technique zones width is related to concentration and not to band broadening mechanisms (see Section 2.15.6).

3.5.1 From Single Bands or Peaks

3.5.1.1 Number of Theoretical Plates

Martin and Synge's article [104] was the first (known) publication to use the concept of *theoretical plates*, from the theory of distillation processes, as a formal measurement of separation efficiency in chromatography. The number of theoretical plates (N) can be calculated either in space or in time. It is defined as:

$$N = \underbrace{\frac{l^2}{\sigma^2(x, t_m)}}_{\text{In space}} \simeq \underbrace{\frac{t_m^2}{\sigma^2(l, t)}}_{\text{In time}}. \tag{3.1}$$

In space: l is the distance run by the center of mass of the analyte band in the reference frame of the platform wall. This is measured from $c(x, t_m)$, which is a snapshot of the molecular distribution along the separation path taken at time t_m (see Section 2.6 and Figure 2.7). This spatial profile of the bands at $t = t_m$ allows spatial variance $\sigma^2(x, t_m)$ to be determined.

In time: A detector is positioned at $x = l$ and the signal (electropherogram with peaks) is registered. This allows the migration time (t_m) of the center of mass of the peak and its variance $\sigma^2(l, t)$ to be determined from $c(l, t)$ (also shown in Section 2.6 and Figure 2.7).

Therefore, $\sigma^2(x, t_m)$ denotes the variance of the band along the separation medium (in space) at time t_m, and $\sigma^2(l, t)$ denotes the variance of the peak in the electropherogram (in time) when a detector is positioned at $x = l$. They are related to each other by $\sigma(x, t_m) \simeq \sigma(l, t)[(\mu_{el} + \mu_{eo})E]$, which is only valid when transit times through the detection point are short compared to t_m ($\sigma(l, t) \ll t_m$).

By making simple substitutions within equation 3.1 and assuming that all other band dispersion mechanisms are negligible compared to molecular diffusion, an expression for N is obtained that is identical for both scenarios: in space and in time. Moreover, it can be expressed in terms of the applied electric field E or applied voltage V (the more simple notation V will henceforth be used instead of ΔV):

$$N = \frac{\mu E l}{2D} = \frac{\mu V l}{2DL}, \tag{3.2}$$

where l is either the distance run by the center of mass of the analyte (in the in space version) or the position of the detector relative to L (in the in time version). L is the total length of the separation medium over which the electrostatic potential difference drops by an amount of V (assuming that $dV/dx = V/L$ is constant), μ is the apparent mobility, and D is the molecular diffusion coefficient. Note that l cannot be larger than L in this open layout ($l \leq L$); however l can be much larger than L in the toroidal layout (as shown in Chapter 4).

Equation 3.2 shows that high electrical mobilities and small diffusion coefficients produce high separation efficiencies. Secondly, values of l that are close to L are better than $l \ll L$. Finally, this equation suggests that either strong fields or high voltages are necessary to achieve high separation efficiencies. However, the high temperature gradients that build up across the separation medium with the use of strong applied fields add additional band broadening mechanisms. Therefore there is an optimal field strength for each instrument, which will give the maximum number of possible plates for a given method. An interesting way to overcome this limitation and produce higher efficiencies when using stronger fields is to mitigate the thermal band broadening mechanism by applying a Poiseuille counter-flow. This is advantageous if the separation medium has a vigorous and symmetrical cooling setup (see Appendix F).

3.5.1.2 Number of Theoretical Plates per Unit Time Squared

The *number of theoretical plates per unit time squared* (plate double-rate or time efficiency) is calculated by dividing the expressions in equation 3.2 by the migration time squared. Assuming molecular diffusion is the only band broadening mechanism and by making some simple calculations, the following is obtained:

$$\mathscr{N} = \frac{N}{t_m^2} = \frac{\mu^3 E^3}{2Dl} = \frac{\mu^3 V^3}{2DlL^3}. \tag{3.3}$$

Note that $\mathscr{N}$ has a cubic dependence on μ, E, and V. This is interesting and means that fast migrating analytes perform better. This must be kept in mind when planning derivatization reactions. For instance, in the free-solution electrophoresis mode, where electrophoresis differences are the only parameters responsible for the separations, labeling reactions that produce highly charged derivatives and a minimal increase in friction coefficients should be preferred. The exact meaning and uses of $\mathscr{N}$ will be discussed at the end of Chapter 5.

Finally, taking the Nernst–Einstein equation of the kinetic theory of diffusion through a liquid with a low Reynolds number, $D = k_B T/f$, and replacing D by $k_B T/f$ in equations 3.2 and 3.3, further insights can be obtained. These insights should be considered with caution because temperature also affects the dynamic viscosity of the buffer, which in turn changes the friction coefficient f. Note that D can by replaced by $k_B T/f$ only if the frictional coefficients of the entity under electrophoretic transport is identical to the frictional coefficient under hydrodynamic transport [12,108].

3.5.1.3 Height Equivalent of a Theoretical Plate

The height equivalent to one theoretical plate (H) or simple plate height is defined as the separation length (l) divided by the plate number and the result is a simple equation:

$$H = \frac{l}{N} = \frac{2D}{\mu E} = \frac{2DL}{\mu V}.$$ (3.4)

The shortest plate heights are desired and they have a dimension of length (μm or mm). To achieve this, again, D should be small and μ should be as high as possible (this flexibility exist by working at full charge or when derivatization reactions are used). Note that only molecular diffusion is considered in equation 3.4 and it is difficult to include the contribution of radial temperature gradients on band dispersion as it depends on the geometry of the separation medium, the BGE, and the instrument used. Nevertheless, the value of E that minimizes H should be found for each method and instrument.

3.5.2 From Two Neighboring Bands or Peaks

Important data can be extracted from two neighboring bands or peaks, even when they are barely resolved, as demonstrated in Figures 3.7A and B. The standard deviation of the band or peak on the left and right can be easily found if the bands or peaks are symmetrical and have a Gaussian profile. First the heights h_1 and h_2 must be measured, and then the half-width (w) at any height (h) can be measured. This allows the standard deviation (σ) to be determined using the equations $\sigma_1 = w[2\ln(h_1/h)]^{-1/2}$ and $\sigma_2 = w[2\ln(h_2/h)]^{-1/2}$. If this gives the same values of σ_1 (σ_2) for different heights then this is an indication that the band/peak has a Gaussian (normal) distribution.

A second way to find the standard deviations of 1 and 2 is to measure the half-width at the inflection point (w_{ip}). This half-width is exactly equal to the standard deviation ($\sigma = w_{ip}$). The height of the inflection point is located at the heights $s_1 = 2h_1 e^{-1/2}$ and $s_2 = 2h_2 e^{-1/2}$. A third way to discover the standard deviation is to measure the half-width of the base (w_b), which is equal to two times the standard deviation ($2\sigma = w_b$). The easiest way to determine this half-width is: make a dot above band 1 along its center of mass line at the point ($\bar{x}_1$, $2h_1 e^{-1/2}$), and then drawing a straight line (dashed lines in Figure 3.7B) from this point and running tangent to the band. The intercept of this straight line (tangent crossing the inflection point) with the horizontal axis defines the half-width (distance from the center of mass to the intercept). The same can be done for band 2. The expressions of these linear equations (dashed lines in Figure 3.7B), which run tangent to s at the inflections points, are: $t_1 = (x - \bar{x}_1 + 2\sigma_1)h_1 e^{-1/2}/\sigma_1$ for band 1 and $t_2 = (-x + \bar{x}_2 + 2\sigma_2)h_2 e^{-1/2}/\sigma_2$ for band 2, where $\bar{x}$ represents the positions of the centers of mass. The variable x should be replaced by t when working with peaks in electropherograms. The standard deviations values produced are exact values for isolated Gaussian bands/peaks, while for partially overlapped

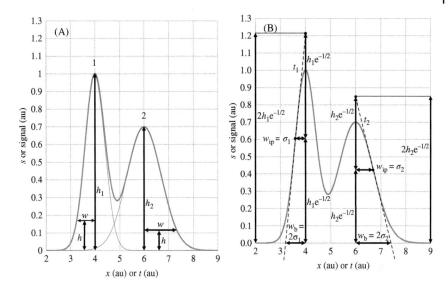

Figure 3.7 Two barely resolved hypothetical bands or peaks (1 and 2). Their approximate standard deviations can be found if they have a Gaussian (normal) distribution. The thick curve represents the band/peak profile, while the thin curves represent the real distributions of bands/peaks 1 and 2. (A) The standard deviation of band/peak 1 can be calculated from many positions of the left side of band/peak 1 using $\sigma_1 = w[2 \ln(h_1/h)]^{-1/2}$, where w is the half-width at any height h and h_1 is the height of the band/peak at maximum. The standard deviation of band/peak 2 can be calculated in the same way using its right side. (B) Alternatively, the standard deviations of bands/peaks 1 and 2 can be calculated by measuring the half-width at the inflection point, as here the half-width (w_{ip}) is equal to the standard deviation. The inflection points of bands bands/peaks 1 and 2 are located at $s = h_1 e^{-1/2}$ and $s = h_2 e^{-1/2}$, respectively. Finally, the half-width at base (w_b), defined as the distance between the center of mass and the intercept of the straight line (dashed line) with the horizontal axis, is equal to two times the standard deviation. These straight lines (hypotenuse of the right-angle triangle) are tangents drawn to the inflection points. Knowing that the heights of these triangles lie at points $s_1 = 2h_1 e^{-1/2}$ and $s_2 = 2h_2 e^{-1/2}$ facilitates the drawing of these tangent lines.

bands/peaks a small error may occur depending on the degree of overlap, heights, and their relative standard deviations.

The first and second derivatives of overlapped bands or peaks are very informative. The distance from the position of the first maximum of the first derivative (first inflection point) to the center of mass of band/peak 1 (approximately equal to its maximum) gives the standard deviation of band/peak 1 (as shown in Figure 3.8A). The distance from the position of the last maximum of the first derivative (last inflection point) to the center of mass of band/peak 2 gives the standard deviation of band/peak 2. The first maximum of the second derivative (3.31) and the last maximum of the second derivative (7.21) are shown in Figure 3.8B.

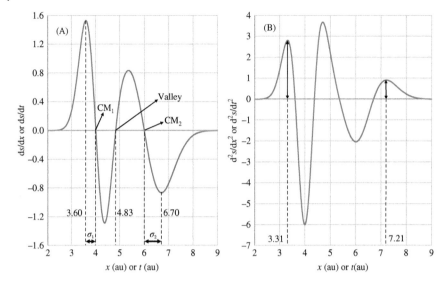

Figure 3.8 First derivatives (A) and second derivatives (B) of two unresolved bands or peaks. (A) The first maximum value of the first derivative (3.60), valley minimum (4.83), last minimum of the first derivative (6.70), and the positions of the centers of mass (CM) are shown. Values of σ_1 and σ_2 can be calculated from this data, from the distances of CM_1 and CM_2 to the external inflection points. (B) The first maximum (3.31) and last maximum (7.21) of the second derivative are shown. The exact values of area, height, first derivative maxima, and second derivative maxima are shown in Tables 3.1 and 3.2. The respective estimated values that would be obtained from the raw data of scanned bands, or peaks on the electropherograms, are also shown in these tables. Note the high precision of the second derivatives (partially overlapped bands/peaks).

Note that these two parameters lie predominately outside the main overlapping region, which is located between the two maxima of band/peaks 1 and 2.

Barely resolved bands and peaks also cause some errors in quantitative analyses, especially when their areas and heights are used to build the calibration curves. Tables 3.1 and 3.2 show the mathematically exact areas (exact) calculated from the thin curves (Gaussians) of Figure 3.8A, the estimated areas measured from

Table 3.1 Parameters extracted from band/peak 1.

	Exact	From raw data	% Error
Area	0.996	0.640	−35.7
Height	0.988	1.000	1.2
M1stD	1.516	1.530	0.9
M2ndD	2.789	2.806	0.6

Table 3.2 Parameters extracted from band/peak 2.

	Exact	From raw data	% Error
Area	1.228	1.186	−3.4
Height	0.700	0.700	<0.01
M1stD	−0.866	−0.866	<0.01
M2ndD	0.911	0.911	<0.01

the raw data (raw data), taken by splitting the two overlapped bands/peaks at the valley minimum, and the respective differences calculated in percentage (% error). The heights, maximum height of the first derivative on the left (M1stD), and the maximum height of the second derivative on the left (M2ndD) are also given. All of the above calculations are repeated for the right hand side band or peak, taking the maximum of the first and second derivatives from its right side. Note the decreasing error in percentage values from **area** (which has large errors), **height**, **M1stD**, and **M2ndD** (which has the smallest error values). In this case, concentration calibration curves made using the M2ndD should give the most precise results, with M1stD giving the second most precise results, followed by height, and finally area.

There are dozens of programs that perform the above mentioned calculations from raw data (spatially scanned bands and electropherograms). Unfortunately only a few of them are open source programs, which allow their codes to be accessed and their calculations checked. Nevertheless, it is easy to measure the parameters shown in Figure 3.8 with a ruler. The parameters shown in Figures 3.8

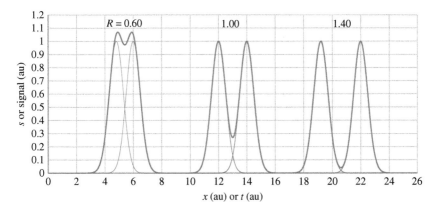

Figure 3.9 Resolution of partially overlapped bands/peaks. The thin curves are the exact profile of the bands/peaks if they have Gaussian distributions. The thick curve is what would be detected by a scanner (bands) or an end point detector (for electropherograms). The exact resolutions are: $R = 0.60$, 1.00, and 1.40.

are also easily calculated; however they require a program or a spreadsheet (e.g., Microsoft Excel™). This also shows that the parameters (standard deviations, heights, and centers of mass) of two or more overlapped distributions can be determined, one-by-one from the outside inwards, in an iterative manner. This is the *modus operandi* of most integration programs and of the dedicated software sub-routines known as peak-callers.

3.5.2.1 Resolution

The resolution $R_{i,j}$ between two neighboring bands or peaks (i and j) is given by:

$$R_{i,j} = \underbrace{\frac{x_i - x_j}{2[\sigma_i(x, t_m) + \sigma_j(x, t_m)]}}_{\text{In space}} = \underbrace{\frac{t_{m,j} - t_{m,i}}{2[\sigma_i(l, t) + \sigma_j(l, t)]}}_{\text{In time}}, \tag{3.5}$$

where x_i and x_j are the positions of the *centers of mass*, and $\sigma_i(x, t_m)$ and $\sigma_j(x, t_m)$ are, respectively, the spatial standard deviations of bands i and j at time t_m. The parameters $t_{m,i}$, $t_{m,j}$, $\sigma_i(l, t)$ and $\sigma_j(l, t)$ are extracted from the peaks of the electropherograms and their migration times and standard deviations are expressed in units of time. It is assumed that $x_i > x_j$ and, consequently, $t_{m,i} < t_{m,j}$.

The resolution, calculated from equation 3.5 when considering $x_i > x_j$ ($t_i < t_j$), gives real and positive numbers that serve as a criterion to decide whether or not two neighboring bands or peaks are resolved or separated. Figure 3.9 shows two Gaussian bands/peaks in three situations: with resolutions of $R = 0.60$, $R = 1.00$, and $R = 1.40$. Any two bands or peaks with $R \geq 1$ are considered fully separated even if the valley between the bands (peaks) does not reach the bottom line. However, if $0 \leq R_{i,j} < 1$ then they are considered unresolved.

Figure 3.10 illustrates the effect of symmetry and the importance of working with the centers of mass of the bands/peaks instead of with their maxima. All three hypothetical separations have resolutions exactly equal to one ($R = 1$). The first pair is made of gaussian (symmetric) bands/peaks, the second pair (the center pair) contains one asymmetric band with its tail towards its neighbor, and the third pair also contains one asymmetrical band, but its tail points away from its neighbor. Using their maxima to calculate their resolutions gives the incorrect results of $R = 1.00$, 1.25, and 0.75, respectively. This example shows the importance of working with the centers of mass instead of the maxima (or modes) of the distributions.

Substituting $\sigma_i^2 = 2D_i t$, $l = \frac{\mu_i + \mu_j}{2} Et$ and $t = (t_{m,i} + t_{m,j})/2$ into equation 3.5, the following very useful equation is obtained:

$$R_{i,j}(t) = \frac{\mu_i - \mu_j}{2(\sqrt{D_i} + \sqrt{D_j})} \left(\frac{Vl}{2\mu L}\right)^{\frac{1}{2}}, \tag{3.6}$$

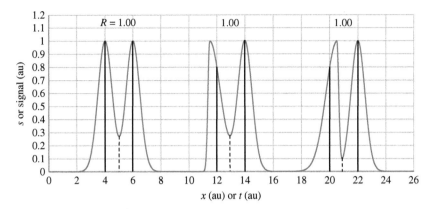

Figure 3.10 The effect of asymmetry on the calculation of resolution between two neighboring bands/peaks. When the center of mass is used to calculate the resolution the correct value of $R = 1.00$ is found for all situations. However, when the band or peak maxima are used the incorrect values of $R = 1.00$, 1.25, and 0.75 (from left to right) are found.

where $\mu_i - \mu_j$ is the apparent *mobility* difference. Equation 3.6 is expected to be valid for AE, ELFSE, FSE, MEEKC, MEKC, and SE in any platform of the open layout. However, the toroidal layout produces different expressions for R and they are shown in Chapter 4. In the case of free-solution electrophoresis, where only the electric force and the viscous friction force act on the analytes, the electrical mobility is equal to the electrophoretic mobility and the apparent mobility difference is given by $\mu_i - \mu_j = \mu_{e,i} - \mu_{e,j} = \frac{q_i}{f_i} - \frac{q_j}{f_j}$. It is also important to mention that $\overline{\mu}$ is the harmonic mean of the *apparent mobilities*, $\overline{\mu} = 2\mu_i\mu_j/(\mu_i + \mu_j)$, and it is not the arithmetic mean as normally stated in the literature. If $\mu_i \simeq \mu_j$, which is the case for hard to separate mixtures, then the harmonic mean approaches the arithmetic mean. Equation 3.6 can be rewritten in a very didactic and explicit form with four terms:

$$
\begin{aligned}
R_{i,j} &= \frac{2^{-\frac{3}{2}}}{\sqrt{D_i} + \sqrt{D_j}}(\mu_i - \mu_j)\left(\frac{1}{\overline{\mu}}\right)^{\frac{1}{2}}(El)^{\frac{1}{2}} \\
&= \frac{2^{-\frac{3}{2}}}{\sqrt{D_i} + \sqrt{D_j}}(\mu_i - \mu_j)\left(\frac{1}{\overline{\mu}}\right)^{\frac{1}{2}}\left(\frac{Vl}{L}\right)^{\frac{1}{2}}.
\end{aligned}
\tag{3.7}
$$

This equation shows how to maximize resolution in the open layout techniques, those whose runs have both a start and an end (or an inlet and an outlet). The first term demonstrates that low molecular diffusion coefficients favor high resolutions. The second term shows something obvious: that mobility differences are very important. In free solution electrophoresis the selection of optimal pH is crucial to ensure maximum mobility difference. For the other electrophoresis modes the proper additives, their respective concentrations, and pH must be optimized. The third term supposes that the average apparent mobility is higher than zero, otherwise meaningless *imaginary numbers* would be obtained for the resolution. However, fine tuning the electroosmotic mobility to make it as close to the average electrical mobility as possible can cause the resolution to spike to an ultra-high value. In this case the runs will take a long time, as the analyte will pass the detection point slowly. Moreover, as pointed out in Section 2.9, electroosmosis is not a reproducible variable and this gives poor repeatabilities among a succession of these long runs. The last term shows that high voltages should be preferred for high resolutions, but overheating degrades the resolution due to the appearance of additional band broadening mechanisms, as discussed in Sections 2.11, 2.12, and 2.13. Therefore, the value of the applied electric field strength that maximizes the resolution must be found. Finally, for the open layouts discussed here, the best location of the detectors is at the end of the separation medium, which gives $l = L$. The rate l/L demonstrates the fraction of V that is effectively used for the separation. If the upper limit of the output of the high voltage source is not a problem then, in this case, l does not need to be maximized as it is enough to increase V to achieve maximum resolutions.

Finally, equations 3.2 and 3.6 allow resolution to be written as the simple expression:

$$R_{ij} = \frac{1}{4}\left(\frac{\mu_i - \mu_j}{\overline{\mu}}\right)N^{\frac{1}{2}}. \tag{3.8}$$

This shows that resolution is proportional to both the square-root of the number of plates, and the ratio of mobility difference over the mean apparent mobility. Finally, it must be mentioned that there are alternative roads (out of the box thinking ways) that may eventually take resolution to new levels in the open layouts. A few of them are based on the use of novel BGE (isoelectric buffers and non-aqueous buffers) and the use of additives (ionic liquids). They are reviewed and studied by Varenne and Descroix [109].

3.5.2.2 Resolution per Unit Time

Equation 3.8 shows the role of the operational parameters in order to get the maximum resolution, regardless of the time it takes to achieve it. Therefore, the expression of resolution per unit time should indicate the operation parameters that are important in order to get the maximum resolution in the shortest run time.

Dividing equation 3.8 by the run time (t) gives:

$$R_t = \frac{R_{i,j}}{t} = \frac{2^{-\frac{3}{2}}}{\sqrt{D_1} + \sqrt{D_2}}(\mu_i - \mu_j)\left(\frac{\overline{\mu}}{l}\right)^{\frac{1}{2}} E^{\frac{3}{2}}$$

$$= \frac{2^{-\frac{3}{2}}}{(\sqrt{D_1} + \sqrt{D_2})}(\mu_i - \mu_j)\left(\frac{\overline{\mu}}{l}\right)^{\frac{1}{2}}\left(\frac{V}{L}\right)^{\frac{3}{2}}. \tag{3.9}$$

Note that R_t is proportional to the applied electric field strength raised to the power 3/2. Moreover, according to equation 3.7, R is inversely proportion to $\sqrt{\overline{\mu}}$, while R_t is directly proportional to $\sqrt{\overline{\mu}}$.

3.5.3 From n Bands and n Peaks

Band and peak capacity is a very useful performance indicator when simultaneously looking at n bands or peaks. However, the discussion must be separated into bands (along the separation track) and peaks (on the electropherogram), as the calculations are very different and more complicated in the latter case. See Section 2.6 for a detailed discussion of the differences between bands and peaks.

3.5.3.1 Band Capacity
Consider the space occupied by a symmetrical band along the separation medium as 4σ, i.e. it starts at 2σ, before the center of mass of the band, and finishes after 2σ. Bands separated in this way exhibit a resolution exactly equal to one ($R = 1$). In this case a rough indication of the maximum number of bands that can be separated in a single run is given by:

$$B = \frac{L}{4\sigma}, \tag{3.10}$$

where L is the total length of the separation medium. Making simple substitutions to this equation produces:

$$B = \frac{1}{4}\left(\frac{\mu E l}{2D}\right)^{\frac{1}{2}} = \frac{1}{4}\left(\frac{\mu V l}{2DL}\right)^{\frac{1}{2}}. \tag{3.11}$$

Note that a high mobility and applied voltage increase band capacity. There is a simple relationship between band capacity and plate number:

$$B = \frac{1}{4}N^{\frac{1}{2}}. \tag{3.12}$$

In conclusion, both resolution (equation 3.8) and band capacity (equation 3.12) are proportional to the square root of the number of theoretical plates. This is the origin of the synonym "separation efficiency" for the number of plates (N).

3.5.3.2 Band Capacity per Unit of Time

A given platform with an associated separation method that produces a large number of bands in a short time interval is much better than another that produces the same number of bands in a long time interval. Therefore, the definition of number of bands per unit time (B_t) is a very interesting parameter and it pays to look at the operating parameters that can optimize this performance indicator.

Dividing equation 3.11 by the run time gives

$$B_t = \frac{B}{t} = \frac{1}{4}\left(\frac{\mu^3 E^3}{2Dl}\right)^{\frac{1}{2}} = \frac{1}{4}\left(\frac{\mu^3 V^3}{2DlL^3}\right)^{\frac{1}{2}}. \tag{3.13}$$

Note the importance of mobility and applied electric field strength, as it appears at the power of 3/2. Small molecular diffusion coefficients are also important. According to this theory there is nothing else that can be done in the open layout to improve the band capacity per unit time.

3.5.3.3 Peak Capacity

There are many equations in the literature to calculate the peak capacity (P) from electropherograms. Nashabeh and El Rassi [111] proposed an expression for electropherograms of free-solution electrophoresis separation of proteins:

$$P = 1 + \frac{\sqrt{N_{av}}}{4}\ln\left(\frac{t_Z}{t_A}\right), \tag{3.14}$$

where N_{av} is the average plate number of the electropherogram and t_A and t_Z are the migration times of, respectively, the first-eluting (A) protein and the last-eluting (Z) protein.

Foley et al. [112] deduced an expression which, among other advantages, maintains the principle of additivity (the sum of the peak capacity of segments of the electropherogram is equal to the peak capacity of the whole electropherogram):

$$P = \frac{l}{2R_S\sqrt{2D}}\left(\frac{1}{\sqrt{t_A}} - \frac{1}{\sqrt{t_Z}}\right), \tag{3.15}$$

where R_S is an dimensionless parameter expressing the desired resolution between peaks (the default is $R_S = 1$). Equations 3.14 and 3.15 are used to calculate the peak capacity of a given method from existing electropherograms. In order to have some guidance in method development it is interesting to have P expressed in terms of the operation parameters such as E, V, μ, μ_{eo}, and l. After some substitutions in equation 3.15 the following was obtained [112]:

$$P = \frac{1}{2R_S}\left(\frac{El}{2D}\right)^{\frac{1}{2}}(\sqrt{\mu_A} - \sqrt{\mu_Z}). \tag{3.16}$$

Considering $R_S = 1$, which is the smallest accepted value for two neighbor peaks to be considered resolved, and rewriting equation 3.16 in a more compact form, results in the following:

$$P = \frac{1-\Gamma}{2}\left(\frac{\mu_A El}{2D}\right)^{\frac{1}{2}}, \tag{3.17}$$

where $\Gamma = \sqrt{\mu_Z/\mu_A}$ is a real number, positive, and smaller than one $(0 < \Gamma < 1)$.

3.5.3.4 Peak Capacity per Unit Time

Dividing equation 3.16 by the migration time of the last peak (t_Z) and after some algebraic simplifications results in the following [112]:

$$P_t = \frac{P}{t} = \frac{1}{2R_S}\left(\frac{E^3}{2Dl}\right)^{\frac{1}{2}}(\sqrt{\mu_A} - \sqrt{\mu_Z})\mu_Z. \tag{3.18}$$

This equation, derived by Foley et al. [112], can also be written in a more compact form. Putting $R_S = 1$ and using the variable Γ previously defined, results in the following:

$$P_t = \frac{P}{t} = \frac{(1-\Gamma)\Gamma}{2}\left(\frac{\mu_A^3 E^3}{2Dl}\right)^{\frac{1}{2}}. \tag{3.19}$$

Note that both P and P_t increase with μ_A and E; however the power dependencies are significantly different (1/2 and 3/2, respectively). By using $l = L$ it is possible to see that P is proportional to $V^{1/2}$ while P_t is proportional to $(E^3/L)^{1/2}$. A graphical analysis of equations 3.16 and 3.18 shows the effects of μ_{eo}, μ_A, and μ_Z on P and P_t [112].

References

1 Hjertén, S. (1967). *Chromatographic Reviews* 9: 122–219.
2 Hjertén, S. (1970). *Methods of Biochemical Analysis* 18: 55–79.
3 Virtanen, R. (1874). *Acta Polytechnica Scandinavica* 123: 1–67.
4 Pretorius, V., Hopkins, B.J., and Schieke, J.D. (1974). *Journal of Chromatography* 99: 23–30.
5 Mikkers, F.E.P., Everaerts, F.M., and Verheggen, Th.P.E.M. (1979). *Journal of Chromatography* 169: 1–10.
6 Mikkers, F.E.P. and Everaerts, F.M. (1979). *Journal of Chromatography* 169: 11–20.
7 Jorgenson, J.W. and Lukacs, K.D. (1981). *Analytical Chemistry* 53: 1298–1302.
8 Jorgenson, J.W. and Lukacs, K.D. (1981). *Journal of High Resolution Chromatography* 4: 230–231.

9 Jorgenson, J.W. and Lukacs, K.D. (1981). *Journal of Chromatography* 218: 209–216.

10 Jorgenson, J.W. and Lukacs, K.D. (1983). *Science* 222: 266–272.

11 Bente, P., Zerenner, E.H., Dandeneau, R.D., and Hewlett-Packard Corp. (1981). US Patent 4, 293, 415.

12 Khaledi, M.G. (ed.) (1998). *High-Performance Capillary Electrophoresis: Theory, Techniques and Applications*. New York: Wiley.

13 Ali, I., Sanagi, M.M., and Aboul-Enein, H.Y. (2014). *Electrophoresis* 35: 926–936.

14 Suntornsuk, L. (2010). *Analytical and Bioanalytical Chemistry* 398: 29–52.

15 Thormann, W., Lurie, I.S., McCord, B. et al. (2001). *Electrophoresis* 22: 4216–4243.

16 Herrero, M., Simo, C., García-Cañas, V. et al. (2010). *Electrophoresis* 31: 2106–2114.

17 Piñero, M.-Y., Bauza, R., and Arce, L. (2011). *Electrophoresis* 32: 1379–1393.

18 Mantovani, V., Galeotti, F., Maccari, F., and Volpi, N. (2018). *Electrophoresis* 39: 179–189.

19 Partyka, J., Krenkova, J., Cmelik, R., and Foret, F. (2018). *Journal of Chromatography A* 1560: 91–96.

20 García, A., Godzien, J., López-Gonzálvez, A., and Barbas, C. (2017). *Bioanalysis* 9: 99–130.

21 Zhang, W., Hankemeier, T., and Ramautar, R. (2017). *Current Opinion in Biotechnology* 43: 1–7.

22 Poinsot, V., Carpéné, M.-A., Bouajila, J. et al. (2012). *Electrophoresis* 33: 14–35.

23 Denoroy, L. and Parrot, S. (2017). *Separation and Purification Reviews* 46: 108–151.

24 Wan, H. and Blomberg, L.G. (2000). *Journal of Chromatography A* 875: 43–88.

25 Scriba, G.K.E. (2003). *Electrophoresis* 24: 4063–4077.

26 Herrero, M., Ibañez, E., and Cifuentes, A. (2008). *Electrophoresis* 29: 2148–2160.

27 Haselberg, R., de Jong, G.J., and Somsen, G.W. (2007). *Journal of Chromatography A* 1159: 81–109.

28 Dawod, M., Arvin, N.E., and Kennedy, R.T. (2017). *Analyst* 142: 1847–1866.

29 Hajba, L. and Guttman, A. (2017). *TrAC Trends in Analytical Chemistry* 90: 38–44.

30 Durney, B.C., Crihfield, C.L., and Holland, L.A. (2015). *Analytical and Bioanalytical Chemistry* 407: 6923–6938.

31 Schmutz, J., Wheeler, J., Grimwood, J. et al. (2004). *Nature* 429: 365–368.

32 Venter, J.C., Adams, M.D., Myers, E.W. et al. (2001). *Science* 291: 1304–1351.

33 Voeten, R.L.C., Ventouri, I.K., Haselberg, R., and Somsen, G.W. (2018). *Analytical Chemistry* 90: 1464–1481.

34 Harrison, D.J., Manz, A., Fan, Z. et al. (1992). *Analytical Chemistry* 64: 1926–1932.

35 Harrison, D.J., Fluri, K., Seiler, K. et al. (1993). *Science* 261: 895–897.

36 Castro, E.R. and Manz, A. (2015). *Journal of Chromatography A* 1382: 66–85.

37 Coltro, W.K.T., Piccin, E., da Silva, J.A.F. et al. (2007). *Lab on a Chip* 7: 931–934.

38 Truckenmüller, R., Rummler, Z., Schaller, T., and Schomburg, W.K. (2002). *Journal of Micromechanical Microengineering* 12: 375–379.

39 Truckenmüller, R. and Giselbrecht, S. (2004). *IEE Proceedings of Nanobiotechnology* 151: 163–166.

40 Adamski, K., Kubicki, W., and Walczak, R. (2016). *Procedia Engineering* 168: 1454–1457.

41 Anciaux, S.K., Geiger, M., and Bowser, M.T. (2016). *Analytical Chemistry* 88: 7675–7682.

42 Cocovi-Solberg, D.J., Worsfold, P.J., and Miró, M. (2018). *TrAC Trends in Analytical Chemistry* 108: 13–22.

43 Sochol, R.D., Sweet, E., Glick, C.C. et al. (2018). *Microelectronic Engineering* 189: 52–68.

44 Wenclawiak, B.W. and Püschl, R.J. (2006). *Analytical Letters* 39: 3–16.

45 da Silva, J.A.F. and do Lago, C.L. (1998). *Analytical Chemistry* 70: 4339–4343.

46 Tanyanyiwa, J., Leuthardt, S., and Hauser, P.C. (2002). *Electrophoresis* 23: 3659–3666.

47 Ghanim, M.H. and Abdullah, M.Z. (2011). *Talanta* 85: 28–34.

48 Hebert, N.E., Kuhr, W.G., and Brazill, A.S. (2002). *Electrophoresis* 23: 3750–3759.

49 Vandaveer, W.R. IV, Pasas-Farmer, S.A., Fischer, D.J. et al. (2004). *Electrophoresis* 25: 3528–3549.

50 Liu, Y., Huang, X., and Ren, J. (2016). *Electrophoresis* 37: 2–18.

51 Anazawa, T., Uchiho, Y., Yokoi, T. et al. (2017). *Lab on a Chip* 17: 2235–2242.

52 Dang, F., Zhang, L., Hagiwara, H. et al. (2003). *Electrophoresis* 24: 714–721.

53 Yang, X., Yan, W., Bai, H. et al. (2012). *Optik* 123: 2126–2130.

54 Llopis, S.D., Stryjewski, W., and Soper, S.A. (2004). *Electrophoresis* 25: 3810–3819.

55 Chun, H., Dennis, P.J., Ferguson Welch, E.R. et al. (2017). *Journal of Chromatography A* 1523: 140–147.

56 Du, Y. and Wang, E. (2007). *Journal of Separation Science* 30: 875–890.

57 He, X., Chen, Q., Zhang, Y., and Lin, J.-M. (2014). *TrAC Trends in Analytical Chemistry* 53: 84–97.

58 Kitagawa, F. and Otsuka, K. (2011). *Journal of Pharmaceutical and Biomedical Analysis* 55: 668–678.

59 Schappler, J., Veuthey, J.-L., and Rudaz, S. (2008). *Separation Science and Technology* 9: 477–521.

60 Salimi-Moosavi, H., Jiang, Y., Lester, L. et al. (2000). *Electrophoresis* 21: 1291–1299.

61 Lu, Q. and Collins, G.E. (2001). *Analyst* 126: 429–432.

62 Ro, K.W., Lim, K., Shim, B.C., and Hahn, J.H. (2005). *Analytical Chemistry* 77: 5160–5166.

63 Collins, G.E., Lu, Q., Wu, N., and Talanta, P. (2007). *Pereira* 72: 301–304.

64 Ohlsson, P.D., Ordeig, O., Mogensen, K.B., and Kutter, J.P. (2009). *Electrophoresis* 30: 4172–4178.

65 Knight, J.B., Vishwanath, A., Brody, J.P., and Austin, R.H. (1998). *Physical Review Letters* 80: 3863–3866.

66 Haswell, S.J. and Skelton, V. (2000). *TrAC Trends in Analytical Chemistry* 19: 389–395.

67 Poe, B.L., Haverstick, D.M., and Landers, J.P. (2012). *Clinical Chemistry* 58: 725–731.

68 Lin, X., Wu, J., Li, H. et al. (2013). *Talanta* 114: 131–137.

69 Marchiarullo, D.J., Sklavounos, A.H., Oh, K. et al. (2013). *Lab on a Chip* 13: 3417–3425.

70 Wuethrich, A. and Quirino, J.P. (2019). *Analytica Chimica Acta* 1045: 42–66.

71 Carrilho, E. (2000). *Electrophoresis* 21: 55–65.

72 Bousse, L., Mouradian, S., Minalla, A. et al. (2001). *Analytical Chemistry* 73: 1207–1212.

73 Štěpánová, S. and Kasicka, V. (2019). *Journal of Separation Science* 42: 398–414.

74 Štěpánová, S. and Kasicka, V. (2016). *Journal of Separation Science* 39: 198–211.

75 Figeys, D. and Pinto, D. (2001). *Electrophoresis* 22: 208–216.

76 Raisi, F., Belgrader, P., Borkholder, D.A. et al. (2001). *Electrophoresis* 22: 2291–2295.

77 Fogarty, B.A., Lacher, N.A., and Lunte, S.M. (2006). *Methods in Molecular Biology* 339: 159–186.

78 Nuchtavorn, N., Suntornsuk, W., Lunte, S.M., and Suntornsuk, L. (2015). *Journal of Pharmaceutical and Biomedical Analysis* 113: 72–96.

79 Guihen, E. and O'Connor, W.T. (2010). *Electrophoresis* 31: 55–64.

80 Yang, M.-P., Huang, Z., Xie, Y., and You, H. (2018). *Chinese Journal of Analytical Chemistry* 46: 631–641.

81 Chagas, C.L.S., Duarte, L.C., Lobo, E.O. et al. (2015). *Electrophoresis* 36: 1837–1844.

82 Munro, N.J., Huang, Z., Finegold, D.N., and Landers, J.P. (2000). *Analytical Chemistry* 72: 2765–2773.

83 Throckmorton, D.J., Shepodd, T.J., and Singh, A.K. (2002). *Analytical Chemistry* 74: 784–789.

84 Wang, J., Chen, G., and Pumera, M. (2003). *Electroanalysis* 15: 862–865.

85 García, C.D. and Henry, C.S. (2003). *Analytical Chemistry* 75: 4778–4783.

86 Kato, M., Gyoten, Y., Sakai-Kato, K., and Toyo'oka, T. (2003). *Journal of Chromatography A* 1013: 183–189.

87 Wang, J., Mannino, S., Camera, C. et al. (2005). *Journal of Chromatography A* 1091: 177–182.

88 Gawron, A.J., Martin, R.S., and Lunte, S.M. (2001). *European Journal of Pharmaceutical Sciences* 14: 1–12.

89 Ferey, L. and Delaunay, N. (2016). *Separation and Purification Reviews* 45: 193–226.

90 Wu, M., Chen, W., Wang, G. et al. (2016). *Food Chemistry* 209: 154–161.

91 Vlčková, M. and Schwarz, M.A. (2007). *Journal of Chromatography A* 1142: 214–221.

92 Kazarian, A.A., Hilder, E.F., and Breadmore, M.C. (2010). *Analyst* 135: 1970–1978.

93 Shimizu, T. and Kenndler, E. (1999) *Electrophoresis* 20: 3364–3372.

94 Aras, N. and Yalçin, S. (2014). *Journal of Analytical Atomic Spectrometry* 29: 545–552.

95 Zarei, M., Zarei, M., and Ghasemabadi, M. (2017). *TrAC Trends in Analytical Chemistry* 86: 56–74.

96 Pomastowski, P. and Buszewski, B. (2014). *TrAC Trends in Analytical Chemistry* 53: 167–177.

97 Rabilloud, T., Chevallet, M., Luche, S., and Lelong, C. (2010). *Journal of Proteomics* 73: 2064–2077.

98 Ünlü, M., Morgan, M.E., and Minden, J.S. (1997). *Electrophoresis* 18: 2071–2077.

99 Minden, J. (2007). *BioTechniques* 43: 739–745.

100 Warnick, G.R., McNamara, J.R., Boggess, C.N. et al. (2006). *Clinics in Laboratory Medicine* 26: 803–846.

101 Lopez-Canovas, L., Martinez Benitez, M.B., Herrera Isidron, J.A., and Flores Soto, E. (2019). *Analytical Biochemistry* 573: 17–29.

102 McKelvey-Martin, V.J., Green, M.H.L., Schmezer, P. et al. (1993). *Mutation Research - Fundamental and Molecular Mechanisms of Mutagenesis* 288: 47–63.

103 Lee, R.F. and Steinert, S. (2003). *Mutation Research – Reviews in Mutation Research* 544: 43–64.

104 Martin, A.J.P. and Synge, R.L.M. (1941). *Biochemical Journal* 35: 1358–1368.

105 Giddings, J.C. (1969). *Separation Science* 4: 181–189.

106 Beale, S.C. and Sudmeier, S.J. (1995). *Analytical Chemistry* 67: 3367–3371.

107 Clark, B.K. and Sepaniak, M.J. (1995). *Journal of Microcolumn Separations* 7: 593–601.

108 Bockris, J.Oâ.M. and Reddy, A.K.N. (1998). *Modern Electrochemistry*, vol. 1, 2e, 505–526. New York: Plenum Press.

109 Varenne, A. and Descroix, S. (2008). *Analytica Chimica Acta* 528: 9–23.

110 Terabe, S., Shibata, O., and Isemura, T. (1991). *Journal of High Resolution Chromatography* 14: 52–55.

111 Nashabeh, W. and El Rassi, Z. (1991). *Journal of Chromatography* 559: 363–383.

112 Foley, J.P., Blackney, D.M., and Ennis, E.J. (2017). *Journal of Chromatography A* 1523: 80–89.

4

Toroidal Layout

4.1   Introduction

Toroidal electrokinetic techniques are characterized by a toroidal or closed loop separation path. As explained in the next sections, microholes are excavated onto the wall of the toroidal capillaries, micro-entrances are fabricated in microchips, and millimeter size conic holes onto the walls of toroidal chambers. They are used as hydrodynamic and electrical connections between the inner lumen of the toroidal paths and the reservoirs containing the buffer solutions. After injection, the sample's components of interest (analytes) are electrodriven in a *quasi*-continuous circulating mode until the desired resolution and/or peak capacity are achieved. This is made by rotating the applied high voltages at certain intervals of time (as explained in Chapter 6).

Similar circulating modes of separation are also used in chromatography in which two or more columns are connected by two or more valves. The outlet flexible stainless steel tube of column 1 is connected to a valve and this is connected to the inlet flexible stainless steel tube of column 2 and vice versa. Driving the substances consecutively through one and then through the other column could, in principle, simulate an infinite length column in this way. Obviously there are dead volumes presented by these flexible stainless steel tubes and valves. Recycling chromatography and closed-loop chromatography are the names used in the literature for these pressure driven, valve controlled, and cyclic circulating modes of chromatographic separations [1].

The absolute minimum number of microholes and reservoirs for an EST with a toroidal layout to work is two. However, this is a very special case because the EOF must be non-zero and, more important than this, one branch must have a positively charged surface (EOF goes from cathode to anode in this branch) and the other branch must have a negative charged surface (EOF goes from anode to cathode in this branch). Finally, the electroosmotic mobility must be larger than the electrophoretic mobility of the faster running analytes, otherwise they will migrate

Open and Toroidal Electrophoresis: Ultra-High Separation Efficiencies in Capillaries, Microchips, and Slabs, First Edition. Tarso B. Ledur Kist.
© 2020 John Wiley & Sons Ltd. Published 2020 by John Wiley & Sons Ltd.

out of this toroidal loop. This was named electrophoretron [2] and for more details see the reference of Eijkel et al. [1].

Toroidal systems containing three, four, and five microholes or microconnections are the most common, but there is no limit for the maximum number that can be used. However, instruments with four equally spaced microholes or microconnections are by far the most advantageous compared to all other possible setups.

The advantages of toroidal systems with four microholes or microconnections are:

(1) Applying vacuum or pressure to two oppositely positioned reservoirs allows the whole toroid to be flushed. This is not possible if three or five are used.

(2) The electric field strength will be homogeneous inside the lumen when an electrostatic potential difference is applied between opposite reservoirs. This is important, for instance, when a dual-opposite injection is used for the simultaneous analysis of cations and anions. Note that this is not possible when an odd number is used.

(3) When an electric potential V is applied to one reservoir and an electric potential $-V$ is applied to the opposite reservoir then the electric potential at the intermediate reservoirs will be approximately zero. The advantages of this are discussed in Section 4.8.

(4) The use of two detectors positioned close to two oppositely positioned microholes or reservoirs presents many interesting advantages. It increases the peak capacity as it prevents the loss of undetected bands to the reservoirs, which may happen when the distance between the back runners and the front runners increases too much.

The mathematical expressions of the performance indicators of the ESTs with a toroidal layout as a function of the operation parameters and with an arbitrary number of m microholes or microconnections were recently calculated [3]. In this book only the best setup, made of four microholes and reservoirs, will be considered. Consequently, a small change in the notation from Kist [3,4] will be adopted here so that the performance indicators of a toroidal system with four microholes can be precisely compared to the open layouts. Simply put, the high electric potential differences will always be applied between opposite reservoirs and the electric potential of the intermediate reservoirs will be left floating, except when the electrodynamic active mode of operation is used (explained in Section 4.8). Moreover, the total length of the toroid will be denoted by $2L$ in this book and not L as in the cited references [3,4]. In this case the expressions of the applied electric field strength will be the same for both the open and toroidal layouts and given by $E = V/L$. This makes the comparison of the performance indicators much more simple and direct. The detector performs two detections per turn: at x close to L (after making almost one-half of a turn with time duration of $\sim \tau$) and at x close to $2L$ (after almost completing a turn with time duration of $\sim 2\tau$), and so on. The samples are injected into the lumen through one of the microholes at $x = 0$ and set to run in a clockwise ($\circlearrowright$) manner. The high voltage distributors, which form an important part of the toroidal ESTs, are discussed in Chapter 6.

4.2 Toroidal Capillary Electrophoresis

The pioneering works of Zhao et al. and Zhao and Jorgenson [5,6] are the precursors of the toroidal layouts of the ESTs using the fused silica capillary platform. They used four pieces of capillary 50 cm in length each, 50 µm ID, and 365 µm OD. The ends were cut flat and polished to ensure a flat and smooth cross sectional surface on both ends. The four capillaries were then connected by sliding the ends through a teflon guide to ensure a perfect lineup at the joins and a loop 2 m long was produced in this way. Moreover, a solenoid was used to move one capillary of each joint so that the gap could be opened and closed. These gaps are always left open (for rinsing, sample introduction, and to work as an electric connection), except when the set of bands crossed the respective position (then it is closed to prevent band leaking).

Fusion splicing is routinely used to joint optical fibers. Some adaptions are required to joint the ends of fused silica microtubes (capillaries) [4]. First the coating (usually poly(imide)) must be removed from the tips; the hot sulfuric acid technique works very well for this [4], although there are many other techniques to accomplish this [7]. Then the tips must be properly cleaved and polished to ensure flat and right angles at these ends [8–10]. Finally, the two ends of a single capillary are fused together [4]. Figure 4.1A shows the two ends of a capillary with 50 µm ID and 360 µm OD positioned and ready for the fusion splicing and Figure 4.1B shows the final product, with the ends permanently jointed by fusion. The two most common techniques to perform this is the filament heating technique and the arc melting technique. The result is a torus (toroid with a cylindrical geometry) as illustrated in Figure 4.2. In this figure the toroidal capillary is bent into a rounded-off rectangle (to minimize the bench footprint) and contains four microholes (which is the most advantageous number of all). Figure 4.3 shows the same toroid but without the poly(imide) coating and folded into a squeezed oval geometry. The best positions for the detectors are shown in this figure, assuming that the sample is injected into the toroid through the microhole positioned inside reservoir one and set to run in a clockwise manner.

Microholes are excavated onto the wall using FIB, laser ablation, or other cutting or etching techniques. The inset of Figure 4.2 shows the shape of an ideal microhole (see Subsection 4.6). During the manufacturing process the reservoirs containing electrodes are installed on top of these microholes. Buffer solutions are

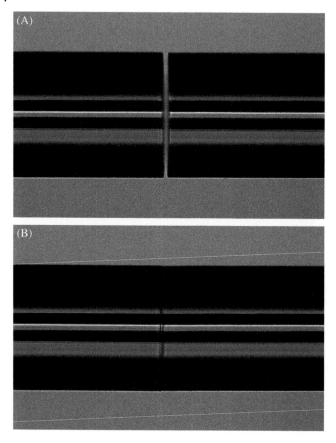

Figure 4.1 A) Photograph of the two ends of a 50 μm ID and 365 μm OD fused silica capillary, which are straight cleaved and polished, and are ready for the fusion splicing. (B) Photograph of the permanently joined ends after the fusion splicing procedure using the filament "furnace" technology. The final product is a torus, which is a toroid with a cylindrical geometry. (*Source*: Courtesy of Vytran, a division of Thorlabs, Inc.)

added to the reservoirs and the inner lumen of the torus is rinsed by applying a vacuum or some pressure to two oppositely positioned reservoirs. The following steps are used to inject the sample: (i) the buffer solution of one reservoir is removed; (ii) a drop of the sample is pipetted onto the top of the microhole (a small secondary reservoir of about 1 mL positioned over this microhole facilitates sample injection); (iii) pressure is applied to this reservoir; (iv) the sample is removed and a buffer is added to this reservoir; (v) the electrostatic potential difference is immediately applied to the two neighboring reservoirs to make the analytes run in the clockwise direction.

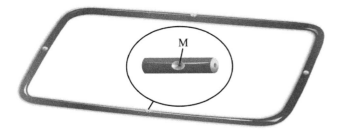

Figure 4.2 Illustration of the toroidal layout of the capillary platform using four equally spaced microholes. This toroid is folded into a rounded-off rectangle with the microholes positioned in the middle of each side. The inset shows the ideal shape of the microholes (discussed in Section 4.6).

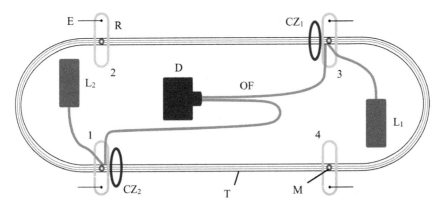

Figure 4.3 Illustration showing the same capillary toroid (T) of Figure 4.2, but now with the external coating (poly(imide)) removed and shaped in an squeezed oval geometry. It also shows the main components, which are: the four microholes (M), four reservoirs (R), four electrodes (E), two band compression zones (CZ1 and CZ2, see Appendix G), and a LIF detection system. This detection system is composed of two lasers (L), optical fibers (OF), and one light detector (D). Spatial scanning detectors can also be used with this bare capillary made toroid. Samples are injected into reservoir 1 and set to run in a clockwise (↻) manner. In this setup the analytes are detected twice per turn, one time before crossing the microhole of reservoir 3 and then before crossing the microhole of reservoir 1.

Many detectors (LIF, C^4D, and absorption) are compatible with this setup, both the scanner type of spatial detectors and the fixed positioned detectors. Figure 4.3 shows a convenient optical fiber based detection system that uses two diode lasers and one light detector. This light detector can be either a photomultiplier tube (PMT), an avalanche photodiode (APD), or a charged-coupled device (CCD) detector.

Toroidal capillaries, which are manufactured by permanently joining the ends of a single piece of a fused silica capillary using fusion splicing [4], present a list of interesting properties that are important to achieve high separation efficiencies:

(1) They have a mirror smooth inner surface.
(2) This surface is chemically homogeneous over its whole extension.
(3) The inner radius has a constant value along the whole axial length of $2L$.
(4) The cylindrical geometry facilitates the rational design of the possible cooling strategies (presented in Chapter 7).
(5) The etched microholes withstand their topography even with repeated rinsing with basic solutions (e.g., 1 M KOH).
(6) There are no connectors, fittings, glues, or moving parts along the torus.

Therefore, the highest separation efficiencies, resolution, and peak capacity of an EST can be expected from the toroidal layout of this fused silica capillary platform.

4.3 Toroidal Microchip Electrophoresis

The toroidal microchip setup is very similar to the open layout studied in Chapter 3, the only difference being the closed loop path and the high voltage distribution (see Chapter 6). As shown in Figure 4.4, everything else is almost identical: the detection system, sample injection, and the microchip manufacturing process.

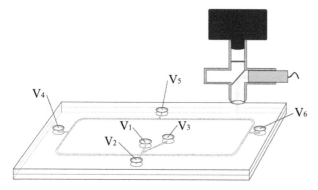

Figure 4.4 Schematic view of the toroidal layout of the microchip platform showing the detector, six reservoirs, and five microconnections between the main toroidal separation path and the lateral microchannels. Reservoirs 1 and 2 are used to inject the sample in the same way as illustrated in Figure 3.4. Reservoirs 3, 4, 5, and 6 contain the running buffer.

The first EST with a toroidal or closed loop layout was demonstrated in a microchip by Burggraff et al. back in 1993 [11,12]. They called it "synchronous cyclic capillary electrophoresis (SSCE)". They used a glass slide where microchannels with a square shape with 2.0 cm side were etched. The channels were 10 μm deep and 40 μm wide. After injection of 100 μM fluorescein and using an electric field strength of 500 V cm^{-1} they noted that the number of theoretical plates increased linearly in time and obtained a total of 40,500 plates in about 6.3 min run time, i.e. after six complete turns. In another article the same authors demonstrated a run that delivered 850,000 plates in 5.9 min. These authors made many improvements and demonstrated also a toroidal microchip electrophoresis using microchannels arranged in a triangle [13–15].

The connections or entrances into main closed loop microchannels play a fundamental role. The entrances of the lateral microchannels into the main microchannel should be kept as small as possible. If they are too wide then band dispersion will be excessively large each time the bands cross the connectors. This is measured by monitoring the variances of the bands at each crossing, as explained in Section 4.8.

4.4   Toroidal Slab Electrophoresis

Toroidal slab electrophoresis is performed by joining the ends of a slab. They may be cast (agarose) or polymerized (acrylamide) into this geometry or they can be bent (to some extend) and inserted into a chamber with a toroidal shape. Figure 4.5A shows a vertical toroidal slab and the respective sample injection device 4.5B. In this case only one sample, injected into the tall and narrow well, can be analyzed at a time. This allows high resolution preparative separations to be performed. Figure 4.6 shows a horizontal toroidal geometry with the wells on top, allowing many samples to be injected and run in parallel. The array of holes (~1 mm in diameter) work as the electric connections between the slab and the buffer solutions in the reservoirs.

4.5   Folding Geometries

It is possible to form many geometries with a closed loop path. For the capillary platform, the most obvious geometry is the ring format. However its bench footprint will be too big if its torus axial length is 1 m or longer. Other options include the oval and the squeezed oval format. The latter contains two long, straight separation paths connected by two tight-radius turns (*many sprint cycling tracks and velodromes have an oval or squeezed oval shape*). There are many other possible

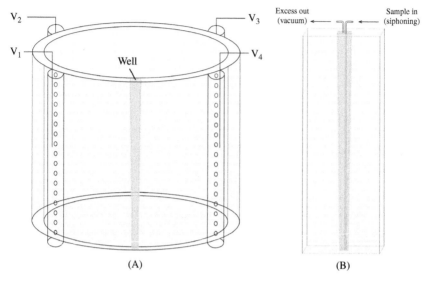

Figure 4.5 Schematic view of the slab platform with a toroidal layout in the vertical version (A) and the respective device required for sample injection into the narrow and tall well (B). The slab and the reservoirs are electrically connected through the line of small holes (~1 mm in diameter).

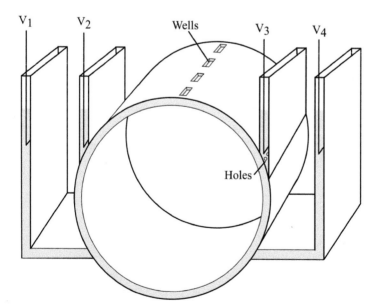

Figure 4.6 Schematic view of the slab platform with a toroidal layout in the horizontal version. The slab and the reservoirs are electrically connected through the line of small (~1 mm) holes.

(A)

(B)

(C)

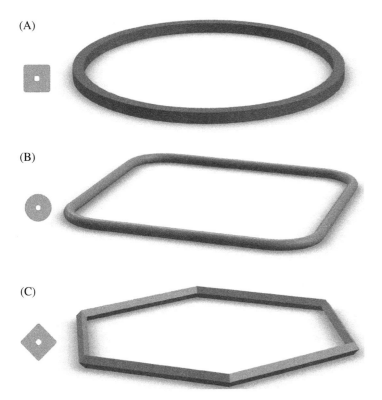

Figure 4.7 Three examples of folding geometries: (A) toroid with a square cross section folded in a ring format. (B) Toroid with a cylindrical cross section folded into a rounded-off square. (C) Toroid with a square cross section folded into an hexagon with sharp edges (instead of rounded-off edges).

geometries such as D-oval, rounded-off square, rounded-off trapezoids, and many rounded-off polyhedrons. Figure 4.7 shows three folding geometries of three toroids with different cross sectional geometries. However, long rounded-off rectangles with a narrow width are the most interesting as they have a narrow bench foot print. Finally, very compact geometries such as toroids coiled into very small solenoids are also possible; however they may cause extra band broadening during the runs in the same way as predicted for open capillary electrophoresis [16].

Many toroids made of flexible fused silica microtubes (capillaries) can be folded into ellipses and overlaid such that many samples can be run in parallel in the same instrument. Figure 4.8 illustrates such an example, showing eight tori, the holders (H), the reservoirs (R), microholes (M), and two detectors (D). Eight samples are injected into reservoirs of holder 1 and set to run in a clockwise manner by applying an electrostatic potential difference between reservoirs 2 and 4. When the analytes are midway between reservoirs 1 and 2 then the application of the

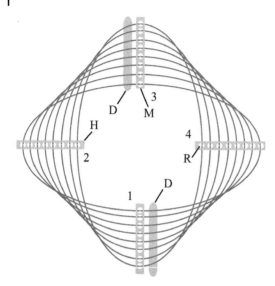

Figure 4.8 Schematic view of eight tori folded as ellipses. They are held in this position by four holders (H) containing eight reservoirs (R) each. The thirty-two microholes (M) are shown and two detectors (D) are used in this set-up (peak capacity is maximized by the use of two detectors).

electrostatic potential is rotated again and applied between reservoirs 1 and 3, and so forth.

In the microchip platform, where microchannels are etched or stamped onto the flat surface of dielectric materials, a zig-zag pattern must be used for the closed loop path in order to fit long tracks into a small chip. This induces band broadening if the microchannels are wide at the tight-radius turns. Moreover, it may even be difficult to etch a meter long microchannel onto a glass slide as good electrical insulation between neighboring microchannels must be maintained. The main advantage of open and toroidal microchip platforms is their compactness compared to capillary and slab platforms. However, as will be shown, the resolution and peak capacity obtained by the microchip platform is expected to be lower than those of the capillary platform when the same separation modes are used.

The number of possible geometries is more limited in the slab platform due to the way it is built. It must have at least one cylindrical or elliptical wall to sustain the slab. The slab can be made of a gel, nylon membrane, cellulose derivative, viscous liquid, or another substance with a low Reynolds number. However, excessively bending thick separation tracks increases band dispersion significantly and therefore reduces separation efficiency. In other words, band dispersion is very strong in coiled running tracks when they are thick [16]. Nevertheless, there are ways to mitigate the band dispersion caused by this mechanism in slabs (see Chapter 7). The main advantage of the slab platforms over the other platforms is the amount of sample that can be processed in preparative separations and the main advantage of the slab platform with a toroidal layout compared to the open layout is the higher separation efficiencies and resolutions that can be expected for these preparative separations.

4.6   Microholes and Connections

Microholes (on capillary walls), micro-entrances (in microchips), and holes (on the slab containing cylinders) are crucial for the performance of the toroidal layouts. They must provide the electrical and hydrodynamic contact between the inner lumen and the external environment, but they must not cause too much band broadening when the active mode of operation is switched on (see Section 4.8). In capillary platforms this can be achieved by excavating the capillary wall to produce microholes with a conic- or funnel-like shape, i.e. shapes with a wide top and a very narrow base (entrance to the lumen). This ideal geometry is shown in Figure 4.9. Note that it does not impose significant electrical resistance on the ionic flow. This is important to prevent the microhole from overheating due to the Joule effect. It also does not increases the overall *hydraulic resistance*, otherwise it would restrict the buffer flow rate during the operation of pressure driven flushing.

Hydraulic resistance (R_h) is an important concept for ESTs and represents the resistance imposed by a conduit to liquid flow in response to a given pressure (ΔP) or level difference (ΔH). This can be written as:

$$R_h = \underbrace{\frac{\Delta P}{\Delta \text{Vol}/\Delta t}}_{\text{Pressure driven}} = \underbrace{\frac{\rho g \Delta H}{\Delta \text{Vol}/\Delta t}}_{\text{Gravity driven}}, \tag{4.1}$$

where $\Delta \text{Vol}/\Delta t$ is the volumetric flow rate of the buffer solution. The parameter R_h is the hydrodynamic resistance, the analog of electrical resistance (R) given by

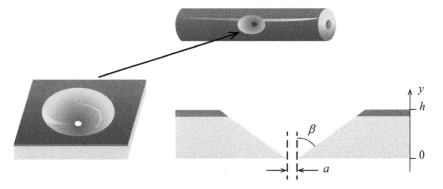

Figure 4.9 Schematic view of the ideal geometry of a microhole etched onto the capillary wall (top) and a perspective view and vertical cut of a simplified flattened version (bottom). *a* denotes the width of the microhole at the entrance into the lumen (separation medium), β is one-half the internal angle of the truncated cone, and *h* denotes the total thickness of the capillary wall (fused silica wall plus coating).

$R = \Delta V/(\Delta Q/\Delta t)$, where ΔV is the elctrostatic potential difference and ΔQ the electric charge.

ESTs with low values of R_h are very sensitive to pressure differences and unleveled reservoirs, which can easily throw bands in one direction or another inside the conduit. Therefore, high R_h values are preferred. R_h varies based on the conduit geometry, conduit length, and dynamic viscosity of the buffer solution.

The electrical resistance created by a conic microhole depends on the depth of the microhole (h) and the inner angle of its funnel-like shape. Assuming a wall with a thickness of h, a flat top, and a flat bottom, its electrical resistance will be given by [4]:

$$R = \frac{4\rho h}{\pi a[a + 2h\tan(\beta)]},\tag{4.2}$$

where a is the diameter of the tip of the truncated cone (where it enters the inner lumen). The ideal value of a depends on the toroid ID. Values of a that are one-fifth of the value of the toroid ID will not impact the hydraulic or electrical resistance of the system if half the internal angle of the truncated cone or microfunnel (β) is larger than 20°. If β assumes its smallest possible value (zero) the microhole is reduced to a cylinder with an internal diameter of a and a depth of h. This is the worst case scenario as both the hydraulic resistance and the electrical resistance imposed by the microhole will be maximized.

When micro-entrances are used as the connections in microchips they should be, approximately, one-fifth of the width of the main microchannel. Alternatively, a double-deep design can be used to attempt to minimize the loss of bands. This is achieved by reducing the depth of the side microchannels to approximately one-fifth of that of the toroidal separation channel [15].

In slab electrophoresis the slab is confined within two concentric dielectric cylinders that function as walls, and an array of holes must be etched onto one of these walls. In this case conic holes are ideal for the same reasons as previously discussed in relation to the capillary platform. Moreover, equation 4.2 can be applied to these conic holes in slab platforms. The necessary number of holes to achieve a good electrical connection depends on the thickness of the slab, with the number of necessary holes increasing with slab thickness. As a rule of thumb, the total area of the entrances of the channels into the lumen ($n\pi a^2$) must be about one-fifth of the cross sectional area of the slab (slab width × height).

4.7 Reservoirs

In toroidal capillary electrophoresis at least one reservoir must be installed over each microhole. There are many ways to do this, and one of them consist of screwing the reservoirs over the microholes with the aid of white rubber O-rings and a dielectric material made hollow screw. This hollow makes the connection between the buffer in the reservoir and the microholes. A second way consists of simply gluing the capillary over a borosilicate glass slide and then gluing the reservoir with a central hole on the bottom over this glass slide containing the capillary with the microhole.

Attention should be taken to the inner volume of the reservoirs. They must be much larger than that used in the open layout because the lifetime of the running electrolytes must be extended for much longer runs. It is true that the electrodes in each reservoir will work alternatively as the cathode and then as the anode. Nevertheless, they must contain a volume of electrolyte that is large enough to withstand for much longer runs.

Instead of using large reservoirs it is possible to use a small reservoir containing many compartments, with the electrode installed into the last compartment [17]. Salt bridges are used to connect the electrode compartment to the neighbor intermediate compartment, the intermediate compartments among themselves, and the first intermediate compartment with the compartment that is in contact with the microhole. Smaller reservoirs can be used with these inner compartments as the electrolysis products, which are generated at the electrodes, are kept away from the solution that is in direct contact with the microholes.

The use of large reservoirs is even more difficult in toroidal microchip electrophoresis due to the lack of space in these compact devices. Therefore, the use of many compartments interconnected with salt bridges is also the best solution for this case. One interesting solution is to build the reservoirs as segments of large and deep channels that are interconnected with short and narrow microchannels that work as salt bridges.

In the case of toroidal slab electrophoresis the ionic currents used to drive the ESTs are in general much larger compared to the capillary and microchip platforms. Therefore in this case tanks containing the buffer solutions with large volumes are required. Alternatively, reservoirs with internal multicomparments can also be used in this case. The connections between these compartments must be compatible with the electric (ionic) current that is planned to be used. Microscopic salt bridges are inappropriate in this case because they may cause overheating in the salt bridges.

4.8 Active and Passive Modes of Operation

In the passive mode of operation no measures are taken to prevent the bands leaking from the inner lumen, through the microholes or connections, to the reservoirs. Moreover, these microholes and connectors have a detrimental effect on the band variances, as they work as dead volumes. Separations are possible in the passive mode [11,12], however part of the bands are lost at each passage at these microholes or connectors. This makes quantitative analyses challenging, as it limits the number of turns that can be performed.

On the other hand, the active mode of operation takes measures to prevent the bands from leaking into the reservoirs. A special accessory is needed to run the active mode. It is selected based on the technique used, and is programmed to run in a cyclic manner. Despite this, in most cases at least a low level of band dispersion will be caused by the active mode of operation. However, band leaking can be completely prevented, bringing enormous advantages to both quantitative analyses and highly efficient separations.

4.8.1 The Gravimetric Method

In the gravimetric method the intermediate reservoir, which the bands cross, is elevated before the arrival of the first band (front runner) and is maintained in this elevated position until the passage of the last band (back runner). This elevation promotes a small inflow of buffer solution from the reservoir to the lumen and to the neighboring reservoirs. This prevents the bands from leaking into the reservoirs. For this reason the geometry of the microhole is critical for the performance of the instrument. Large microholes require a larger buffer flux to prevent band leakage, while small microholes require a much smaller buffer flux to prevent band leakage [4]. The contribution of this gravimetric active mode of operation (due to unleveled reservoirs) to band broadening is already known in the literature [18] and the respective time-dependent dispersion coefficient is given by:

$$D_{pvf} = \frac{a^2 \, \bar{v}^2}{48 \, D} = \frac{a^2 \, v_{max}^2}{192 \, D}, \tag{4.3}$$

where a is the cylinder inner radius, D_{pvf} is the time dependent dispersion coefficient due to the "Poiseuille velocity flow", $\bar{v}$ is the average velocity of the buffer induced by the unleveled reservoirs, v_{max} is the maximum velocity of the buffer (this is observed at the center line of the cylinder), and D is the molecular diffusion coefficient. Note that $\bar{v} = v_{max}/2$ and the expression of v_{max} is given by equation 2.15. However, here the distance between reservoirs (microholes) is $L/2$, therefore the variable L of equation 2.15 must be replaced by $L/2$. Putting all terms in a single equation results in the following:

$$D_{pvf} = \frac{a^4 \, \rho \, g \, \Delta H}{384 \, \eta \, L \, D}, \tag{4.4}$$

where a is the inner radius of the microtube, ρ is the density of the fluid, g is the local gravity acceleration, and ΔH is the liquid levels height difference between the raised reservoir (higher) and the opposite reservoir (lower).

4.8.2 The Hydrodynamic Method

Applying a small pressure to the intermediate reservoir gives the same result as obtained by the elevation of the reservoir in the previous subsection. The advantage of this setup is that the valves that must be turned on and off in a cyclic manner are its only moving parts, facilitating the automation of this method. In this case the application of a pressure ΔP creates a buffer flow with v_{max}. Equation 2.14 gives the relationship between ΔP and v_{max}. Joining the terms again results in:

$$D_{pvf} = \frac{a^4 \, \Delta P}{384 \, \eta \, L \, D}. \tag{4.5}$$

Note that here the distance between reservoirs is $L/2$, therefore the variable L of equation 2.14 was also replaced by $L/2$.

4.8.3 The Electrokinetic Method

Applying a small current to the intermediate reservoir is one way of preventing band leaking. The direction of the ionic current depends on the nature of the ions that migrate through the lumen at the position of the intermediate microhole. If cations are going to cross this position then a positive current must be injected. However, if anions are going to cross this position then a current with the opposite sign must be injected. This is achieved by using sources of electric current instead of sources of electric voltage. In this method the band variance will experience a sudden increase because the electric field strength after the microhole will be stronger than it is before the microhole. The standard deviation of the band will change in the following way:

$$\sigma_{after} = \frac{v_{after}}{v_{before}} \sigma_{before} = \frac{E_{after}}{E_{before}} \sigma_{before}. \tag{4.6}$$

Suppose that the electrophoresis applied high voltage is V, the electrical resistance of segment L is R, and that the small current injected into the lumen through the intermediate microhole is i. In this case the sudden variance change that will be observed due to injection of the current i, which prevents band leaking, is given by:

$$\Delta \sigma^2 = \sigma_{before}^2 \left[\left(\frac{V + Ri}{V - Ri} \right)^2 - 1 \right], \tag{4.7}$$

where $Ri \ll V$ and there are no non-linear effects influencing the procedure.

The bands are suddenly dispersed by the amount given by equation 4.7 each time they cross a microhole where the active electrokinetic operation mode is used to prevent band leaking. This is very different to the gravimetric and hydrodynamic operation modes. These modes create a time dependent band dispersion mechanism that acts on the bands ($\sigma^2 = 2\,D_{pvf}\,t$) in addition to the other additive band broadening mechanisms [19], as long as ΔH or ΔP are non-zero. A more complicated situation is created when ΔH and ΔP are simultaneously non-zero. In this case the term ΔP of equation 4.5 must be replaced by $|\Delta P + \rho\,g\,\Delta H|$, where the vertical bars stand for modulus.

4.8.4 Using Microvalves or Microcaps

Microvalves and microcaps (to cover the microholes) can be produced using laser ablation and many etching techniques. These microfabricated accessories can be used to close and open the microholes in a cyclic manner, preventing in this way the band leaking from the torus lumen out into the reservoirs. However, the operation mechanism must fit into the reservoir and the materials used must be such to withstand the alkaline solutions used to flush the capillaries. Moreover, they must be compatible with the high voltage that is periodically applied to the reservoirs. Finally, all these must be achieved without dead volumes, otherwise band dispersion will still occur at each pass through the position of the microvalve or microhole containing a microcap. Once these technological challenges are overcome, then the greatest separation efficiencies can be expected from toroidal electrophoresis with these devices. This is because in this case the microholes will not cause any band dispersion, thus providing a number of theoretical plates per unit length equal to open electrophoresis, which has no microholes or microconnections between the inlet and outlet.

4.9 Performance Indicators for Toroidal Layouts

The performance indicators of the above closed loop layouts will be studied in the next sections and compared with the open layouts in Chapter 5. The performance indicators are: plate number (N), plate number delivered per unit time squared ($\mathcal{N}$), height equivalent of a plate (H), resolution (R), resolution per unit time (R_t), band capacity (B), band capacity per unit time (B_t), peak capacity (P), and peak capacity per unit time (P_t). Attention will be given to the equations expressing these performance indicators as a function of the operation variables. These variables are: applied electric field strength (E), applied voltage (V), total length of the toroid ($2L$), analyte apparent mobility (μ), and analyte molecular diffusion (D). It is expected that these equations will be applicable to all platforms (capillary, microchip, and slab) with a toroidal layout and the majority of their separation modes, including: affinity electrophoresis (AE), end-labeled free-solution electrophoresis (ELFSE), free-solution electrophoresis (FSE), micellar electrokinetic chromatography (MEKC), micro-emulsion electrokinetic chromatography (MEEKC), and sieving electrophoresis (SE).

4.9.1 From Single Bands or Peaks

In this subsection the expressions of performance indicators from single bands or peaks will be calculated when no cyclic band compression events are applied, regardless of whether sample stacking is used or not at the beginning of the runs. Note that on-column band compression can be performed anywhere and either in a cyclic or non-cyclic manner, but this is different from the stacking events that are performed at the beginning of the runs (as explained in Section 2.14). On-line band compression requires hardware and a segment of the capillary where they are performed. Figure 4.3 shows two such segments and they are indicated by CZ_1 and CZ_2. The toroidal layout with cyclic band compression is treated only in Appendix G.

4.9.1.1 Number of Theoretical Plates

The number of theoretical plates is defined as the square of the total distance run by an analyte (L_{tot}^2) divided by its variance σ^2. Suppose that the time taken by the analyte to reach the first detection point (at distance L from the injection point) is τ, therefore, the total distance run by the analyte can be expressed as $L_{tot} = nL = \mu E(n\tau)$. This gives:

$$N = \frac{L_{tot}^2}{\sigma^2} = n\frac{\mu EL}{2D} = n\frac{\mu V}{2D}. \tag{4.8}$$

In this calculation it was assumed that all other band broadening mechanisms are negligible compared to molecular diffusion ($\sigma^2 = 2Dt$) and/or the variance from all other time dependent band broadening mechanisms are additive and, in this case, D represents the sum of all these dispersion mechanisms [19]. Note that the number of plates grows linearly with time ($t = n\tau$), and the number of detections (n), or the number of turns performed by the analyte in the toroidal or closed loop path ($n/2$).

4.9.1.2 Number of Plates per Unit Time Squared

An instrument and its associated method that delivers a given number of plates in a 5 min run is much better than an instrument and its method that takes 50 min to deliver the same number of plates. Moreover, taking the number of plates delivered per unit time squared (plate double-rate or time efficiency) is even more interesting as this variable simplifies simultaneously many important performance indicators (this is explained at the end of Chapter 5). It is given by dividing equation 4.8 by the run time squared:

$$\mathcal{N} = \frac{N}{t_m^2} = \frac{\mu^3 E^3}{2nDL} = \frac{\mu^3 V^3}{2nDL^4}. \tag{4.9}$$

This shows that analytes with high mobilities and low diffusion coefficients perform better. Note that this is almost identical to the expression for the open layout given by equation 3.3, except for the operational parameter n.

4.9.1.3 Height Equivalent of a Theoretical Plate

The height equivalent of a theoretical plate (H) is commonly used in chromatography and, for completeness, will also be calculated. In ESTs the plate height is defined as the ratio of the total distance run by the analyte (L_{tot}) divided by the number of plates (N):

$$H = \frac{L_{tot}}{N} = \frac{2D}{\mu E} = \frac{2DL}{\mu V}. \tag{4.10}$$

Note that these expressions are identical to the open layouts if the detectors of Chapter 3 are positioned at the end of the separation medium (when $l = L$ for the open layouts).

4.9.2 From Two Neighboring Bands or Peaks

4.9.2.1 Resolution

The distance between the centers of mass of two neighboring bands grows as $\Delta x = (\mu_i - \mu_j)Et$. Total run time ($t$) can be expressed as $t = n\tau$, where n is the successive detection that happens when $(x_i + x_j)/2 = nL$ (in space) or when $(t_i + t_j)/2 = n\tau$ (in time). Resolution (R) is defined as $R_{i,j} = \Delta x/2(\sigma_i + \sigma_j)$. Considering that all other band broadening mechanisms are negligible compared to molecular diffusion, the following is obtained:

$$R_{i,j} = \frac{\Delta\mu}{2(\sqrt{D_i} + \sqrt{D_j})} \left(\frac{nEL}{2\bar{\mu}}\right)^{1/2}$$

$$= \frac{\Delta\mu}{2(\sqrt{D_i} + \sqrt{D_j})} \left(\frac{nV}{2\bar{\mu}}\right)^{1/2}, \tag{4.11}$$

where $\Delta\mu = \mu_i - \mu_j$. Moreover, if $t = (t_i + t_j)/2$ then $\bar{\mu} = 2\mu_i\mu_j/(\mu_i + \mu_j)$, or conversely, if $\bar{\mu} = (\mu_i + \mu_j)/2$ then $t = 2t_it_j/(t_i + t_j)$. It is not possible to have $t = (t_i + t_j)/2$ and $\bar{\mu} = (\mu_i + \mu_j)/2$. Note that resolution grows as $n^{1/2}$, however it cannot grow indefinitely because after a while the bands will spread and vanish into the baseline noise, making them undetectable. This is the main advantage of using cyclic band compression, discussed in Appendix G, as it acts against band dispersion. For small analytes with high diffusion coefficients and no cyclic band compression, it takes more than twelve hours for the bands to vanish into the baseline noise. Long run times are not ideal, therefore the resolution generated per unit time is a useful parameter to take into account.

4.9.2.2 Resolution per Unit Time

Dividing equation 4.11 by t and using $t = n\tau$ and $\bar{\mu}E(n\tau) = nL$ gives the following expressions for the resolution per unit time:

$$\frac{R_{i,j}}{t} = \frac{\Delta\mu}{2(\sqrt{D_i} + \sqrt{D_j})} \left(\frac{\bar{\mu}E^3}{2nL}\right)^{1/2}$$

$$= \frac{\Delta\mu}{2L^2(\sqrt{D_i} + \sqrt{D_j})} \left(\frac{\bar{\mu}V^3}{2n}\right)^{1/2}. \tag{4.12}$$

This can be written in a more compact form using $2\sqrt{\overline{D}} = \sqrt{D_1} + \sqrt{D_2}$. Note that in this case $\sqrt{\overline{D}}$ is a notation and not an average. This abbreviated and compact notation is used in the summary Tables 5.2, 5.3, and 5.4.

4.9.3 From n Bands or n Peaks

4.9.3.1 Band Capacity

Band capacity (B) is a performance indicator that indicates the maximum number of analytes, given as bands, that can be separated in a single run. Basically, at each time $n\tau$ the whole arm L of the toroidal path is scanned (toroidal capillary or microchip) or photographed (toroidal slab). Assuming that all analytes have the same diffusion coefficient, then at time $m\tau$ the total number of bands that fit along segment L with $R = 1$ is given by:

$$B_m = \frac{L}{4\sigma} = \frac{L}{4\sqrt{2Dm\tau}} = \frac{1}{4}\left(\frac{\mu EL}{2Dm}\right)^{1/2}. \tag{4.13}$$

The maximum number of bands after n detections, from $m = 1$ to $m = n$, is given by:

$$B = \frac{1}{4}\left(\frac{\mu EL}{2D}\right)^{1/2} \sum_{m=1}^{n} \frac{1}{\sqrt{m}}. \tag{4.14}$$

Calculating the sum of the right hand side of the above equation gives the following final result:

$$B \simeq \frac{(2\sqrt{n}-1)}{4}\left(\frac{\mu EL}{2D}\right)^{1/2}$$

$$\simeq \frac{(2\sqrt{n}-1)}{4}\left(\frac{\mu V}{2D}\right)^{1/2}. \tag{4.15}$$

In the calculation of B it is assumed that at time $t = \tau$ the detector scans all bands from $x = 0$ to $x = L$, at time $t = 2\tau$ it scans from $x = L$ to $x = 2L$, at time $t = 3\tau$ from $x = 0$ to $x = L$ again, and so on. However, these is not always doable because some detectors cannot detect at the position of the reservoirs at $x = 0$, $L/2$, L, and $3L/2$. Note that τ is the time taken by the group of unresolved peaks to run a distance L. Such an example is illustrated by Figure 2.23, that contains a crowd of unresolved peaks as front runners. Therefore, each half of one turn reveals B_m bands (given by equation 4.13) from the crowd, giving a total of B_n after n detections. This is theoretically absolutely the maximum number of bands that can be detected after n detections or $n/2$ turns (or round trips of the front runner). Note that B increases only as $\sqrt{n}$. This occurs due to band broadening, caused by molecular diffusion, which successively limits the number of bands that fit into a segment of length L.

4.9.3.2 Band Capacity per Unit Time

The number of bands that can be separated in a given time interval is an interesting parameter to study because very long runs are costly and annoying in day-to-day laboratory routines. Dividing the expression given by equation 4.15 by $n\tau$, the time needed to separate B bands, the following is obtained:

$$B_t \simeq \left(\frac{1}{2\sqrt{n}}-\frac{1}{4n}\right)\left(\frac{\mu^3 E^3}{2DL}\right)^{1/2}$$

$$\simeq \left(\frac{1}{2\sqrt{n}}-\frac{1}{4n}\right)\frac{1}{L^2}\left(\frac{\mu^3 V^3}{2D}\right)^{1/2}. \tag{4.16}$$

These expressions show that the maximum number of peaks produced in a time interval of $n\tau$ is inversely proportional to $\sqrt{n}$. This happens as band broadening successively limits the number of bands that fit within L, and this in turn decreases the average $B/n\tau$ for increasing values of n. Note that both apparent mobility and electric field strength (and voltage) have the power of $3/2$.

4.9.3.3 Peak Capacity

Peak capacity (P) indicates the number of analytes, given as peaks in the electropherogram that results from the detection of the bands crossing the finishing line, that can be separated in a single run. Although, predicting these peaks seems like a simple job, it is sometimes difficult to find an expression for P as a function of the operation parameters. Equation 3.17 tells us what is produced by a toroid arm of length L. Using this and the effect of the steady increase of the standard deviation ($\propto \sqrt{t}$), which decreases the number of peaks detected after each half turn, produces the following expression:

$$P = \frac{(2\sqrt{n} - 1)(1 - \Gamma_n)}{4} \left(\frac{\mu_{A,n} EL}{2D} \right)^{\frac{1}{2}},$$ (4.17)

where $\Gamma_n = \sqrt{\mu_{Z,n}/\mu_A}$. This is the absolute maximum value of peaks that can be detected after n detections or $n/2$ complete turns in a toroidal capillary, microchip, or slab with axial length $2L$.

To understand the band capacity and peak capacity of toroidal electrophoresis it must be remembered that the group of unresolved front running bands (peaks) have a mobility that goes asymptotically to μ_A for increasing n. As a consequence, $\mu_{A,1} \simeq \mu_{A,2} \simeq \mu_{A,3} = \ldots = L/(E\tau)$. This means that the index n can be removed from $\mu_{A,n}$. However, the index n cannot be removed from $\mu_{Z,n}$ because the mobility of the last detected peak will increase at at each detection, but will be never higher than μ_A. The final result is given by:

$$\begin{aligned} P &= \frac{(2\sqrt{n} - 1)(1 - \Gamma_n)}{4} \left(\frac{\mu_A EL}{2D} \right)^{\frac{1}{2}} \\ &= \frac{(2\sqrt{n} - 1)(1 - \Gamma_n)}{4} \left(\frac{\mu_A V}{2D} \right)^{\frac{1}{2}}. \end{aligned}$$ (4.18)

4.9.3.4 Peak Capacity per Unit Time
Dividing equation 4.17 by $n\tau$ (the time taken to produce the number of peaks given by equation 4.17), results in the following:

$$\begin{aligned} P_t &= \left(\frac{2}{\sqrt{n}} - \frac{1}{n} \right) \frac{\Gamma_n - 1}{2} \left(\frac{\mu_A^3 E^3}{2DL} \right)^{\frac{1}{2}} \\ &= \left(\frac{2}{\sqrt{n}} - \frac{1}{n} \right) \frac{\Gamma_n - 1}{2L^2} \left(\frac{\mu_A^3 V^3}{2D} \right)^{\frac{1}{2}}. \end{aligned}$$ (4.19)

This is the absolute maximum number of peaks produced per unit time. Note that P_t (the rate of peak production) decreases with increasing n. This occurs because the standard deviation of each peak increases with time.

References

1 Eijkel, J.C.T, van den Berg, A., and Manz, A. (2004). *Electrophoresis* 25: 243–252.

2 Choi, J.G., Kim, M., Dadoo, R, and Zare, R.N. (2001). *Journal of Chromatography A* 924: 53–58.

3 Kist, T.B.L. (2018). *Journal of Separation Science* 41: 2640–2650.

4 Kist, T.B.L. (2017). *Journal of Separation Science* 40: 4619–4627.

5 Zhao, J., Hooker, T., and Jorgenson, J.W. (1999). *Journal of Microcolumn Separations* 11: 431–437.

6 Zhao, J. and Jorgenson, J. W. (1999). *Journal of Microcolumn Separations* 11: 439–449.

7 Macomber, J. and Walker, K. (2006). *LC-GC North America* 24: 66.

8 Roeraade, J. (1983). *Journal of High Resolution Chromatography* 6: 140–144.

9 Macomber, J., Hintz, R., Ewing, T. et al. (2005). *LC-GC North America* 23: 81.

10 Macomber, J., Hintz, R., Ewing, T., and Acuña, R. (2005). *LC-GC North America* 23: 90.

11 Burggraf, N., Manz, A., de Rooij, N.F., and Widmer, H.M. (1993). *Analytical Methods and Instrumentation* 1: 55–59.

12 Burggraf, N., Manz, A., Effenhauser, C.S. et al. (1993). *Journal of High Resolution Chromatography* 16: 594–596.

13 Burggraf, N., Manz, A., Verpoorte, E. et al. (1994). *Sensors and Actuators B: Chemical* 20: 103–110.

14 van Heeren, F.V., Verpoorte, E., Manz, A., and Thormann, W. (1996). *Analytical Chemistry* 68: 2044–2053.

15 Manz, A., Bousse, L., Chow, A. et al. (2001). *Fresenius Journal of Analytical Chemistry* 371: 195–201.

16 Kašička, V., Prusík, Z., Gaš, B., and Štědrý, M. (1995). *Electrophoresis* 16: 2034–2038.

17 de Jesus, D.P., Brito-Neto, J.G.A., Richter, E.M. et al. (2005). *Analytical Chemistry* 77: 607–614.

18 Sonke, J.E, Furbish, D.J., and Salters, V.J.M. (2003). *Journal of Chromatography A* 1015: 205–218.

19 Virtanen, R. (1993). *Electrophoresis* 14: 1266–1270.

20 Mandaji, M., Rübensam, G., Hoff, R.B. et al. (2009). *Electrophoresis* 30: 1501–1509.

21 Mandaji, M., Rübensam, G., Hoff, R.B. et al. (2009). *Electrophoresis* 30: 1510–1515.

5

Confronting Performance Indicators

5.1 Introduction

In this chapter the performance indicators of ESTs with open (common) layouts are compared with ESTs with toroidal layouts. They are expected to be valid for the three platforms (capillary, microchip, and slab) and at least the following six separation modes: affinity electrophoresis, end-labeled free-solution electrophoresis, free-solution electrophoresis, micellar electrokinetic chromatography, micro-emulsion electrokinetic chromatography, and sieving electrophoresis.

5.2 Performance Indicators from Experimental Data

Performance indicators are important means for the evaluation of an instrument and method of analysis. Objective performance indicators are, for example: N, number of theoretical plates (plate number or separation efficiency); N_t, number of theoretical plates delivered per unit time (plate rate); $\mathcal{N}$, number of theoretical plates delivered per unit time squared (plate double-rate or time efficiency); H, height equivalent of a theoretical plate (plate height); R, resolution among two neighbor bands or peaks; R_t, resolution produced per unit time; B, band capacity; B_t, band capacity per unit time (band rate); P, peak capacity; P_t, peak capacity per unit time (peak rate).

Many of the above terms are already harmonized and this is reported in many publications such as the IUPAC's 2012 Compendium of Chemical Terminology – Gold Book [1], the IUPAC's 2003 Terminology for Analytical Capillary Electromigration Techniques [2], and the pioneering work of Knox [3]. There are also many important books published in the field with important contributions to the terms and definitions currently used [4–6]. These terms are used as much as possible in this book; however, a few more terms and definitions are necessary. Examples of such additional terms are: (1) number of theoretical plates delivered

Open and Toroidal Electrophoresis: Ultra-High Separation Efficiencies in Capillaries, Microchips, and Slabs, First Edition. Tarso B. Ledur Kist.
© 2020 John Wiley & Sons Ltd. Published 2020 by John Wiley & Sons Ltd.

Table 5.1 Performance indicators (PIs) and the procedure used to calculate them from experimental data (separations). Usually these variables (σ, t_m) are extracted from electropherograms and some times from the results of spatial scans or from photograph taken with the aid of transilluminators. The PIs calculated in these ways are expected to be valid for the two layouts (open and toroidal), three platforms (capillary, microchip, and slab), and at least six separation modes (AE, ELFSE, FSE, MEKC, MEEKC, and SE).

PI	Procedure	Note
$N =$	$\dfrac{l^2}{\sigma^2(x)}$	Number of plates from spatial scans or photograph
	$\dfrac{t_m^2}{\sigma^2(t)}$	Number of plates from electropherograms
$N_t =$	$\dfrac{N}{t_m}$	Number of plates delivered per unit time
$\mathscr{N} =$	$\dfrac{N}{t_m^2}$	Number of plates delivered per unit time squared
$H =$	$\dfrac{l}{N}$	Height equivalent of a theoretical plate
$R =$	$\dfrac{x_1 - x_2}{2[\sigma_1(x) + \sigma_2(x)]}$	Resolution from spatial scans or photograph
	$\dfrac{t_2 - t_1}{2[\sigma_1(t) + \sigma_2(t)]}$	Resolution from electropherograms
$R_t =$	$\dfrac{R}{t_m}$	t_m is the instant of the scan or photograph
	$\dfrac{R}{t_m}$	$t_m = (t_1 + t_2)/2$ is usually used in electropherograms
$B =$	Counting bands	Counted from spatial scans or photograph
$B_t =$	$\dfrac{B}{t_m}$	Bands produced per unit time
$P =$	Counting peaks	Counted from electropherograms
$P_t =$	$\dfrac{P}{t_Z - t_A}$	Peaks produced per unit time

per run time squared (plate double-rate or time efficiency) and represented by $\mathcal{N}$ (this calligraphic N is given by LaTex code \mathcal{N}); (2) the distinctions between B, B_t, P, and P_t; and (3) the nomenclature of some ESTs (detailed in Appendix A).

Table 5.1 summarizes how these performance indicators are calculated from experimental data. The specific experimental data used to calculate them are the following: distances run by the analytes, migration times of the analytes, migration times of the first and last peaks (t_A and t_Z), standard deviations (σ_1, σ_2, ...) of the bands (detected by spatial scanners) or the peaks (given by electropherograms), number of bands (B), and number of peaks (P).

Note that the distance run by the analytes will be the same for all (l) if electropherograms are used. Conversely, the migration times (t_m) will be the same for all if spatial scanners or photographic cameras are used to detect the spatial profile of the bands at time t_m. The variances are the most difficult variable to be extracted from the electropherograms and scans. Figures 3.7 and 3.8 explain a few simple techniques to determine them from bands and peaks.

The mathematical expressions shown in Table 5.1 are extracted from the experimental data (separations) and are useful for quantitative comparisons of the performances of instruments and methods. However, they do not show how the operational parameters should be set in order to maximize these performance indicators. Therefore, expressing these performance indicators in terms of the operational parameters will produce valuable equations that are very useful for both method development and theoretical modeling.

5.3 Performance Indicators Predicted from Operational Parameters

The performance indicators may also be expressed in terms of the operational parameters, such as the applied electric field strength (E), applied voltage (V), apparent mobility (μ), length (l and L), molecular diffusion coefficient (D), and number of detections (n) or the number of turns ($n/2$) performed in the toroidal layout. Although, apparent mobility and diffusion coefficient can only be operated (changed) indirectly (using derivatization reactions, viscosity changes, temperature, and others). The variable n is highlighted in blue in the following to ease the visualization of its role. The goal of the next sections is to find how these performance indicators can be maximized by properly adjusting the cited operational parameters.

These performance indicators were calculated in Chapter 3 for the open layout and in Chapter 4 for the toroidal layout. They are tackled in Tables 5.2, 5.3, 5.4, and 5.5. In Table 5.2 they are listed in terms of applied electric field strength (E), in Table 5.3 in terms of applied voltage (V), in Table 5.4 they are expressed without

Table 5.2 Summary of the performance indicators of the open and toroidal layouts in terms of applied electric field strength (E). They are expected to be valid for the three platforms (capillary, microchip, and slab) and at least six separation modes (AE, ELFSE, FSE, MEKC, MEEKC, and SE)

	Open layout	Toroidal layout
$N =$	$\dfrac{\mu El}{2D}$	$\dfrac{n\mu EL}{2D}$
$\mathcal{N} =$	$\dfrac{\mu^3 E^3}{2Dl}$	$\dfrac{\mu^3 E^3}{2nDL}$
$H =$	$\dfrac{2D}{\mu E}$	$\dfrac{2D}{\mu E}$
$R =$	$\dfrac{\Delta\mu}{4}\left(\dfrac{El}{2\bar{\mu}\bar{D}}\right)^{\frac{1}{2}}$	$\dfrac{\Delta\mu}{4}\left(\dfrac{nEL}{2\bar{\mu}\bar{D}}\right)^{\frac{1}{2}}$
$R_t =$	$\dfrac{\Delta\mu}{4}\left(\dfrac{\bar{\mu}E^3}{2\bar{D}l}\right)^{\frac{1}{2}}$	$\dfrac{\Delta\mu}{4}\left(\dfrac{\bar{\mu}E^3}{2n\bar{D}L}\right)^{\frac{1}{2}}$
$B =$	$\dfrac{1}{4}\left(\dfrac{\mu El}{2D}\right)^{\frac{1}{2}}$	$\dfrac{2\sqrt{n}-1}{4}\left(\dfrac{\mu EL}{2D}\right)^{\frac{1}{2}}$
$B_t =$	$\dfrac{1}{4}\left(\dfrac{\mu^3 E^3}{2Dl}\right)^{\frac{1}{2}}$	$\dfrac{1}{4}\left(\dfrac{2}{\sqrt{n}}-\dfrac{1}{n}\right)\left(\dfrac{\mu^3 E^3}{2DL}\right)^{\frac{1}{2}}$
$P =$	$\dfrac{1-\Gamma}{2}\left(\dfrac{\mu_A El}{2D}\right)^{\frac{1}{2}}$	$\left(2\sqrt{n}-1\right)\dfrac{1-\Gamma_n}{2}\left(\dfrac{\mu_A EL}{2D}\right)^{\frac{1}{2}}$
$P_t =$	$\dfrac{(1-\Gamma)\Gamma^2}{2}\left(\dfrac{\mu_A^3 E^3}{2Dl}\right)^{\frac{1}{2}}$	$\left(\dfrac{2}{\sqrt{n}}-\dfrac{1}{n}\right)\dfrac{(1-\Gamma_n)\Gamma_n^2}{2}\left(\dfrac{\mu_A^3 E^3}{2DL}\right)^{\frac{1}{2}}$

the variable L, whenever possible, and in Table 5.5 they are expressed in terms of plate number (N) and plate double-rate ($\mathcal{N}$). It will be shown that Tables 5.4 and 5.5 are the most important of all, as they give a good panoramic view of how to optimize the performances of the ESTs.

As shown in Table 5.2 the number of plates (N) grows linearly with n and plate double-rate ($\mathcal{N}$) is inversely proportional to n in the toroidal layout (toroidal). However, the expression of plate height is the same for the open and toroidal layouts. The resolution (R) produced by the toroidal layout is proportional to $\sqrt{n}$. The resolution produced per unit time (R_t) is identical for the two layouts if $n = 1$; however, in the toroidal layout it grows only as $1/\sqrt{n}$ for > 1. Band capacity (B) is identical for the open and toroidal layouts if $n = 1$ and $l = L$, then it grows as $\sqrt{n}$ in the toroidal layout if $n > 1$. Band capacity per unit time (B_t) is identical for both layouts if $n = 1$ and $l = L$. Nonetheless, the larger the value of n the smaller

Table 5.3 Summary of the performance indicators in terms of applied voltage (V).

	Open layout	Toroidal layout
$N =$	$\dfrac{\mu V l}{2DL}$	$\dfrac{n\mu V}{2D}$
$\mathcal{N} =$	$\dfrac{\mu^3 V^3}{2DlL^3}$	$\dfrac{\mu^3 V^3}{2nDL^4}$
$H =$	$\dfrac{2DL}{\mu V}$	$\dfrac{2DL}{\mu V}$
$R =$	$\dfrac{\Delta\mu}{4}\left(\dfrac{Vl}{2\bar\mu DL}\right)^{\frac{1}{2}}$	$\dfrac{\Delta\mu}{4}\left(\dfrac{nV}{2\bar\mu D}\right)^{\frac{1}{2}}$
$R_t =$	$\dfrac{\Delta\mu}{4}\left(\dfrac{\bar\mu V^3}{2\bar D lL^3}\right)^{\frac{1}{2}}$	$\dfrac{\Delta\mu}{4L^2}\left(\dfrac{\bar\mu V^3}{2n\bar D}\right)^{\frac{1}{2}}$
$B =$	$\dfrac{1}{4}\left(\dfrac{\mu V l}{2DL}\right)^{\frac{1}{2}}$	$\dfrac{2\sqrt{n}-1}{4}\left(\dfrac{\mu V}{2D}\right)^{\frac{1}{2}}$
$B_t =$	$\dfrac{1}{4}\left(\dfrac{\mu^3 V^3}{2DlL^3}\right)^{\frac{1}{2}}$	$\dfrac{1}{4L^2}\left(\dfrac{2}{\sqrt{n}}-\dfrac{1}{n}\right)\left(\dfrac{\mu^3 V^3}{2D}\right)^{\frac{1}{2}}$
$P =$	$\dfrac{1-\Gamma}{2}\left(\dfrac{\mu_A V l}{2DL}\right)^{\frac{1}{2}}$	$\left(2\sqrt{n}-1\right)\dfrac{1-\Gamma_n}{2}\left(\dfrac{\mu_A V}{2D}\right)^{\frac{1}{2}}$
$P_t =$	$\dfrac{(1-\Gamma)\Gamma^2}{2}\left(\dfrac{\mu_A^3 V^3}{2DlL^3}\right)^{\frac{1}{2}}$	$\left(\dfrac{2}{\sqrt{n}}-\dfrac{1}{n}\right)\dfrac{(1-\Gamma_n)\Gamma_n^2}{2L^2}\left(\dfrac{\mu_A^3 V^3}{2D}\right)^{\frac{1}{2}}$

the band capacity per unit time will be in the toroidal layout. This occurs since the standard deviation of the bands increases with the square root of time, which allows successively fewer bands to fit in the length L.

Notation used in Tables 5.2, 5.3, 5.4, 5.5:

(i) **Open layout**: open layouts with effective separation length l and total length L ($E = V/L$).

(ii) **Toroidal layout**: toroidal layouts (closed-loop separation paths) with four microholes, microconnections, or holes (four reservoirs) and total axial length of $2L$. The high voltage is always applied between opposite reservoirs. In this case, E is also given by $E = V/L$.

(iii) μ_A and $\mu_{Z,n}$ are the apparent mobilities (with respect to the platform wall) of the first and last peak of interest of the nth detection, respectively.

(iv) The operational parameter n represents the number of detections or half-turns. The number of complete turns is given by $n/2$.

(v) The parameter $\sqrt{\overline{D}}$ is a symbol (not the square root of a mean value). Here is what it represents: $\sqrt{\overline{D}} \equiv (\sqrt{D_i} + \sqrt{D_j})/2$.

(vi) $\overline{\mu}$ is the harmonic mean, $\overline{\mu} = 2\mu_i\mu_j/(\mu_i + \mu_j)$, and not the arithmetic mean since $t_m = (t_i + t_j)/2$ was used.

(vii) $\Gamma = \sqrt{\mu_Z/\mu_A}$ and $\Gamma_n = \sqrt{\mu_{Z,n}/\mu_A}$.

Tables 5.2 and 5.3 are a bit strict in the sense that the first uses only E and the latter uses only V. A much better panoramic view is given when only the terms of Tables 5.2 and 5.3 that do not contain the variable L (whenever possible) are used. These more informative expressions are shown in Table 5.4. This is probably the most important table of this chapter as it shows clearly the role of E and V in the optimization of the performance indicators.

Table 5.4 Summary of the nine performance indicators (PI) without the operational parameter L (however the fraction l/L was kept where it appears). This shows when V is important and when E is important. These equations are expected to be valid for the three platforms (capillary, microchip, and slab) and at least six separation modes (AE, ELFSE, FSE, MEKC, MEEKC, and SE).

	PI	Open layout [a]	Toroidal layout [b]
1	$N =$	$\dfrac{\mu Vl}{2DL}$	$\dfrac{n\mu V}{2D}$
2	$\mathcal{N} =$	$\dfrac{\mu^3 E^3}{2Dl}$	$\dfrac{\mu^3 E^3}{2nDL}$
3	$H =$	$\dfrac{2D}{\mu E}$	$\dfrac{2D}{\mu E}$
4	$R =$	$\dfrac{\Delta\mu}{4}\left(\dfrac{Vl}{2\overline{\mu}DL}\right)^{\frac{1}{2}}$	$\dfrac{\Delta\mu}{4}\left(\dfrac{nV}{2\overline{\mu}D}\right)^{\frac{1}{2}}$
5	$R_t =$	$\dfrac{\Delta\mu}{4}\left(\dfrac{\overline{\mu}E^3}{2Dl}\right)^{\frac{1}{2}}$	$\dfrac{\Delta\mu}{4}\left(\dfrac{\overline{\mu}E^3}{2nDL}\right)^{\frac{1}{2}}$
6	$B =$	$\dfrac{1}{4}\left(\dfrac{\mu Vl}{2DL}\right)^{\frac{1}{2}}$	$\dfrac{2\sqrt{n}-1}{4}\left(\dfrac{\mu V}{2D}\right)^{\frac{1}{2}}$
7	$B_t =$	$\dfrac{1}{4}\left(\dfrac{\mu^3 E^3}{2Dl}\right)^{\frac{1}{2}}$	$\dfrac{1}{4}\left(\dfrac{2}{\sqrt{n}}-\dfrac{1}{n}\right)\left(\dfrac{\mu^3 E^3}{2DL}\right)^{\frac{1}{2}}$
8	$P =$	$\dfrac{1-\Gamma}{2}\left(\dfrac{\mu_A Vl}{2DL}\right)^{\frac{1}{2}}$	$\left(2\sqrt{n}-1\right)\dfrac{1-\Gamma_n}{2}\left(\dfrac{\mu_A V}{2D}\right)^{\frac{1}{2}}$
9	$P_t =$	$\dfrac{(1-\Gamma)\Gamma^2}{2}\left(\dfrac{\mu_A^3 E^3}{2Dl}\right)^{\frac{1}{2}}$	$\left(\dfrac{2}{\sqrt{n}}-\dfrac{1}{n}\right)\dfrac{(1-\Gamma_n)\Gamma_n^2}{2}\left(\dfrac{\mu_A^3 E^3}{2DL}\right)^{\frac{1}{2}}$

Table 5.5 Summary of the performance indicators expressed in terms of the number of theoretical plates (N) or number of theoretical plates per run time squared ($\mathcal{N}$). They are expected to be valid for the three platforms (capillary, microchip, and slab) and at least six separation modes (AE, ELFSE, FSE, MEKC, MEEKC, and SE).

		Open layout [a]	Toroidal layout [b]
1	$N =$	$\dfrac{\mu Vl}{2DL}$	$\dfrac{n\mu V}{2D}$
2	$\mathcal{N} =$	$\dfrac{\mu^3 E^3}{2Dl}$	$\dfrac{\mu^3 E^3}{2nDL}$
3	$H =$	$\dfrac{2D}{\mu E}$	$\dfrac{2D}{\mu E}$
4	$R =$	$\dfrac{\Delta\mu}{4\bar{\mu}}\sqrt{N}$	$\dfrac{\Delta\mu}{4\bar{\mu}}\sqrt{N_n}$
5	$R_t =$	$\dfrac{\Delta\mu}{4\bar{\mu}}\sqrt{\mathcal{N}}$	$\dfrac{\Delta\mu}{4\bar{\mu}}\sqrt{\mathcal{N}_n}$
6	$B =$	$\dfrac{1}{4}\sqrt{N}$	$\dfrac{1}{2}\left(1-\dfrac{1}{2\sqrt{n}}\right)\sqrt{N_n}$
7	$B_t =$	$\dfrac{1}{4}\sqrt{\mathcal{N}}$	$\dfrac{1}{2}\left(1-\dfrac{1}{2\sqrt{n}}\right)\sqrt{\mathcal{N}_n}$
8	$P =$	$\dfrac{1}{2}(1-\Gamma)\sqrt{N}$	$\left(1-\dfrac{1}{2\sqrt{n}}\right)(1-\Gamma_n)\sqrt{N_n}$
9	$P_t =$	$\dfrac{1}{2}(1-\Gamma)\Gamma^2\sqrt{\mathcal{N}}$	$\left(1-\dfrac{1}{2\sqrt{n}}\right)(1-\Gamma_n)\Gamma_n^2\sqrt{\mathcal{N}_n}$

Equations 1a and 3a of Table 5.4 were borrowed from the theory of distillation (also a separation method) and introduced into chromatography by Martin and Synger [7]. Later, Giddings [8] introduced them into the field of ESTs along with the definition of resolution (4a), peak capacity (8a), and plate number per unit time or plate rate (N_t). However, the definition of plate number per unit time squared or plate double-rate ($\mathcal{N} = N/t^2$), which is proposed here, is also very useful as shown in Table 5.5. In that pioneering and important work, Giddings substituted D by RT/f instead of $k_B T/f$ because he dealt with a mole of solute species, where R is the gas constant ($R = N_A k_B$), f the friction coefficient, k_B is the Boltzmann constant, and N_A is Avogadro's constant. Note that the substitution $D = k_B T/f$ is valid only when the frictional coefficients involved in electrophoretic transport are identical to the frictional coefficients involved in hydrodynamic transport [6,9,10]. Equation 4b was proposed by Henley and Jorgenson [11] and equations 5a and 7a are a natural consequence of the definitions of resolution and band capacity. Equations 8a and 9a were deduced by Foley et al. [12]. Equation 1b was demonstrated by the pioneering work of Burggraf et al. (figure 3) [13]. To the best of our knowledge, equation 2a along with equations 2b, 5b, 6b, 7b, 8b, and 9b are proposed for the first time here.

Many important conclusions can be drawn from Table 5.4. For instance, the term l/L, which appears in 1a, 4a, 6a, and 8a, has a very clear meaning; it indicates the fraction of the applied high voltage (V) that is effectively used in the separations. Therefore, placing the detector as close as possible to the outlet will improve the performance indicators. The terms nV (in 1b) and $\sqrt{nV}$ (in 4b, 6b, and 8b) indicate that using $V = 30$ kV and $n = 10$ in the toroidal layout would be equivalent to using a high voltage source of $V = 300$ kV in the open layout, assuming that the dispersion mechanisms observed in both layouts are the same.

Table 5.4 also shows that the applied electric potential difference (V) is what matters when the run time is mathematically not important. This is the case for the performance indicators N, R, B, and P. Unfortunately, in this case, the run times may be very long and the performance indicators only grow as the square root of V, i.e. they are proportional only to $V^{1/2}$. This means that the high voltage must be increased by a factor of four to double the performance indicators. Of course, there are other important operational parameters involved, such as n, μ, l, L, $\Delta\mu$, and D, which appear in almost all expressions. On the other hand, the electric field strength (E) is what really matters (and not V) when the output speed of the desired quantities (resolution, bands, and peaks) is important. They are proportional to $E^{3/2}$ (R_t, B_t, and P_t), but limited by thermal dispersion caused by the Joule effect. Some are proportional to $(E^3/l)^{1/2}$ or $(E^3/L)^{1/2}$, which is the case for R_t, B_t, and P_t. This means that if E is already at its maximum then l or L must be decreased in order to improve even further the respective performance indicators. Detection may be challenging in this case as the injected sample plug must be very narrow and the data acquisition rate must be very high to detect the peaks in short run time intervals. Solutions to this may be to apply the flow-gating injection technique [14–18] or to work with the frontal technique (see Section 2.6). In conclusion, there must be a compromise between V and E such that the adequate band capacity or peak capacity (which depends on the V used) is obtained for a given application and in the shortest run time (which depends on E, but limited by Joule heating). The role of the fraction μ/D, which appears in almost all performance indicators, is discussed in Chapter 8.

Comparing the performance indicators of the open and toroidal layout shown in Table 5.4 brings many conclusions. Firstly, the toroidal layout performs better in the separation of hard to separate pairs, for instance stereoisomers, isotopomers, antibody–drug couplings, and isotope profile analyses, to cite a few [11]. This is possible because of the term n that appears in expression 4b. There are additional classes of isomers, such as isomeric peptides, topoisomers, atropisomers, and akamptisomers that require methods with ultra-high resolution for their separation. (*The number of new classes of challenging pairs and mixtures is still growing!*). Secondly, the toroidal layout performs better at separating complex mixtures, such as proteins and protein digests that exhibit similar sizes and mobilities [11]. This is a result of the combined effect of expressions 4b and 6b or 8b (depending on the detection system used). Thirdly, the ELFSE separation mode, where the hard to separate compounds are among the front runners, is also very suitable for the toroidal layout. In this case detecting the bands (expression 6b) would be much better than detecting peaks (expression 8b), because it does not need to wait for the last peak to arrive at the detector before rotating the high voltage (see Chapter 6). Finally, the open layout still performs better at separating complex mixtures if they are made by "well-behaved" analytes such as nucleic acids that display predictable mobilities that are determined mainly by the length of the polyions. In this case the mobility can be adjusted in such a way to maximize both resolution and peak capacity in the runs that are limited by the effective length l. Figure 2.29 is an almost perfect illustration of such an example. For a review of ultra-high resolutions in capillary electrophoresis see Henley and Jorgenson [11].

Finally, it is very interesting to note that all performance indicators can be succinctly expressed in terms of N (plate number or separation efficiency) or $\mathcal{N}$ (plate double-rate or separation efficiency per run time squared). This is shown in Table 5.5.

Table 5.5 can be deduced from Tables 5.2, 5.3, or 5.4 and the result is very simple. The variable N that appears in the expressions of B (P) refers to the plate number of the front running band (peak) and the variable $\mathcal{N}$ that appears in the expression of B_t (P_t) refers to the plate number per migration time squared of the front running band (peak). Another property is that the same term $(1 - \frac{1}{2\sqrt{n}})$ appears in four expressions related to the toroidal layout. Representing this term, for instance, by G_n, would turn the expressions even more compact. Moreover, making $n = 1$ in each expression of the toroidal layout produces the exact corresponding expression of the open layout. In conclusion, the plate number (N) is related to the performance indicators for which the run time is mathematically not important (R, B, and P), while the plate double-rate ($\mathcal{N}$) is related to the performance indicators for which the run time (speed of production) is mathematically important (R_t, B_t, and P_t). To the best of our knowledge, this also is the first record of expressions 5a, 5b, 6b, 7a, 7b, 8b, 9a, and 9b in the literature. Finally, there is an unexpected symmetry among these equations shown in Table 5.5; this is noticeable even by looking at them only at a glance.

References

1 IUPAC (2012). Compendium of Chemical Terminology, Gold Book, Version 2.3.2. Research Triangle Park, NC: International Union of Pure and Applied Chemistry.

2 (a) IUPAC (2003). Terminology for Analytical Capillary Electromigration Techniques. Research Triangle Park, NC: International Union of Pure and Applied Chemistry; (b) Riekkola, M.-L., Jönsson, J.Å., and Smith, A.R.M. (2004). *Pure and Applied Chemistry* 76: 443–451.

3 Knox, J.H. (1994). *Journal of Chromatography A* 680: 3–13.

4 Camilleri, P. (ed.) (1998). *Capillary Electrophoresis – Theory and Practice*, 2e. Boca Raton, FL: CRC Press.

5 Baker, D.E. (1995). *Capillary Electrophoresis*. New York: Wiley.

6 Khaledi, M.G. (ed.) (1998). *High-performance Capillary Electrophoresis: Theory, Techniques and Applications*. New York: Wiley.

7 Martin, A.J.P. and Synge, R.L.M. (1941). *Biochemical Journal* 35: 1358–1368.

8 Giddings, J.C. (1969). *Separation Science* 4: 181–189.

9 Bockris, J.O'M. and Reddy, A.K.N. (1998). *Modern Electrochemistry*, vol. 1, 2e, 505–526. New York: Plenum Press.

10 Stellwagen, E. and Stellwagen, N.C. (2002). *Electrophoresis* 23: 2794–2803.

11 Henley, W.H. and Jorgenson, J.W. (2007). Extreme resolution in capillary electrophoresis: UHVCE, FCCE, and SCCE. In: *Handbook of Capillary and Microchip Electrophoresis and Associated Microtechniques*, 3e (ed. J.P. Lander). Boca Raton, FL: CRC Press. 723–760.

12 Foley, J.P., Blackney, D.M., and Ennis, E.J. (2017). *Journal of Chromatography A* 1523: 80–89.

13 Burggraf, N., Manz, A., Effenhauser, C.S. et al. (1993). *Journal of High Resolution Chromatography* 16: 594–596.

14 Hooker, T.F. and Jorgenson, J.W. (1997). *Analytical Chemistry* 69: 4134–4142.

15 Weng, Q., Fu, L., Li, X. et al. (2015). *Analytica Chimica Acta* 857: 46–52.

16 Zhu, Q., Zhang, Q., Zhang, N., and Gong, M. (2017). *Analytica Chimica Acta* 978: 55–60.

17 Opekar, F. and Tůma, P. (2017). *Journal of Separation Science* 40: 3138–3143.

18 Opekar, F. and Tůma, P. (2019). *Electrophoresis* 40: 587–590.

6

High Voltage Modules and Distributors

6.1 Introduction

There are important differences between the high voltage sources and their connections used in open electrodriven layouts and in toroidal layouts. Therefore, this chapter gives a detailed description of these differences as well as the advantages and disadvantages of each setup.

ESTs are electrodriven by the application of high voltages. The high voltages used typically range from 5 kV to 30 kV in the capillary platform, from 1 kV to 5 kV in the microchip platform, and from 100 V to 1 kV in the slab platform.[1,2] Currents in the slab platforms may reach a few mA, while in the capillary and microchip platforms they rarely cross the 100 μA barrier. The high voltage modules used in the capillary and microchip platforms are based on the Cockcroft–Walton voltage multiplier scheme, while those used in slab platforms are simpler and based on transformers and rectifier bridges. The different ways of connecting the high voltage outputs and the electrodes in the reservoirs are discussed in the following sections for both open and toroidal layouts of the ESTs.

6.2 High Voltages in Open Layouts

Open layouts are characterized by runs with both a start and an end, or an inlet and an outlet. In this case the polarities of the electrodes do not change during the run, except when the pulsed field electrophoresis mode is used.

For safety reasons, the most common setup of wires applies the ground to the inlet electrode and the high voltage to the outlet electrode. In some instruments, however, the opposite is observed. High voltage modules usually have the option of remote polarity control, meaning that the polarity of the high voltage can be changed. Consequently, the inlet reservoir will act as a cathode (lower potential)

Open and Toroidal Electrophoresis: Ultra-High Separation Efficiencies in Capillaries, Microchips, and Slabs, First Edition. Tarso B. Ledur Kist.
© 2020 John Wiley & Sons Ltd. Published 2020 by John Wiley & Sons Ltd.

in some methods while in others it will act as an anode (higher potential). This is important because it allows the two types of analyses to be performed: from cathode to the anode and from anode to the cathode.

A third, rarely used mode of operation involves the application of a regulated positive voltage (from 0 to V) to an electrode and a regulated negative voltage (from 0 to $-V$) to the other electrode. The only advantage of this setup is that two commercially available, simple high voltage modules with outputs of V and $-V$ allow a given EST to be electrodriven with an electrostatic potential difference of $2V$. For instance, using a module at -30 kV and another at 30 kV (both common options on the market) gives an electrostatic potential difference of 60 kV. Note that it is not common to find a single high voltage module with a regulated output from 0 to 60 kV. Many performance indicators are significantly increased by the application of ultra-high voltages, as explained in Chapter 5, and this has been demonstrated in a series of applications using the capillary platform.[3–7]

A very special situation is observed when pulsed field electrophoresis is used. In this case high voltage is applied as a rectangular pulse of a given polarity and duration, followed by a either a null potential for a given time interval or a pulse of the opposite sign. This is commonly used in slab platforms to separate very long, charged polymers (e.g., nucleic acids) and it is rarely used in the capillary and microchip platforms. Moreover, in some applications the direction of the electric field changes with each pulse. For example, if the first pulse is applied in one direction, the second pulse is applied at an angle of 120° from the direction of the first pulse. Next, the third pulse is applied at 240° from the direction of the first pulse, the fourth pulse is applied in the same direction as the first pulse, and so on. This is achieved by standing arrays of electrodes on every side of a hexagonal slab.[8–10]

6.3 High Voltages in Toroidal Layouts

6.3.1 The Ideal Toroidal Length

An important question is: What is the ideal axial length of a toroid that should be used in an EST with a toroidal layout? Another way to ask the same question is: Which toroid length optimizes the performance indicators of Tables 5.4 and 5.5? According to the notations adopted in this book, this axial length is denoted by $2L$.

Tables 5.2–5.5 contain the mathematical equations of important performance indicators, expressed as functions of the operation parameters such as V, E, L, μ, μ_{eo}, and others. The first thing that catches the eye is the importance of the applied voltage (V) for some applications and the applied electric field strength (E) for others. This is very clear in Table 5.4. They play an important role in the maximization of the respective performance indicators. However, there is a limit to the field strengths that can be applied. Above certain values of E the thermally induced band broadening mechanism starts to significantly increase the variances of the bands, which in turn decreases the values of the performance indicators. Secondly, in most cases a small band broadening effect is observed each time a band crosses a microhole (in toroidal capillaries), a microconnection (in toroidal microchannels in microchips) or a hole (in a toroidal slab). Therefore, especially for long runs, the fewer times that a band crosses a microhole/microconnection/hole the better. Consequently, there are two answer to the above questions: (1) The length $2L$ will be defined by the strongest electric field that will be used in the instrument. (2) The toroid should be as long as possible while respecting the limits of the highest output of the commercially available high voltage modules. This ensures that the desired strongest field strength will be reached. Note that this minimizes the number of microholes crossed by a band set to run a given distance.

The electric field strength within a toroid with four reservoirs (and four microholes/microconnections/holes) will be given by $E = V/L$, since the high voltage is always applied between opposite reservoirs separated by a distance of L. In microchips space is always a limitation, which complicates projects using very long microchannels; for instance those with a length of around 1 m. However, this limitation does not exist in the capillary platform where any torus longer than 20 cm can be fabricated and used. Therefore, it is important to define the optimal axial length of a torus that will generate the best performance indicators (given in Chapter 4). High electric field strengths cannot be applied to slab platforms, because of the thickness of the slab and consequently the limited surface-to-volume ratio in this case, which limits the heat dissipation. The advantage of slab platforms is that they can be applied to preparative separations. Therefore, the following discussion will be concentrated on capillary platforms that exhibit the potential to push the performance indicators to their maximum possible values.

In conclusion, the ideal length of a capillary with a toroidal layout can be decided from the following two operational parameters: (1) The strongest electric field that will be applied. This could be around $1000 \, \text{V cm}^{-1}$ [11] or even higher if the Poiseuille counter-flow strategy is used (discussed in Chapter 7 and Appendix F). (2) The maximum output of the commonly used, regulated high voltage modules. These modules must be accessible, convenient, safe, and reliable. The regulated 0 to 30 kV and 0 to −30 kV modules, which have been on the market for a few decades, meet these requirements.

There is still one more point to be analyzed before the ideal toroidal length can be chosen. There are at least three ways of applying an electrostatic potential (V) between two electrodes: (1) Applying the ground to an electrode and V to the opposite electrode. (2) Applying the ground to an electrode and $-V$ to the opposite electrode. (3) Applying $-V$ to an electrode and V to the opposite electrode. Options 1 and 2 are not ideal for many reasons. One such reason is that they will always generate an electric potential ($V/2$ or $-V/2$) in the intermediate reservoirs whose electrodes are left floating when the gravimetric or hydrodynamic active modes of operation are used to prevent band leaking (see Section 4.8). These electric potentials may lead to extra band leaking as the electric current passes through the microholes (because it is difficult to ensure a completely sealed reservoir). If the electrokinetic active mode of operation is used then a small current must be applied to these intermediate reservoirs, which is easier when they float at around zero rather than at approximately $+V/2$ or $-V/2$. Therefore, the ideal situation is to drive the toroidal EST by applying a positive voltage (here called V_+) to an electrode and a negative voltage ($V_- = -V_+$) to the opposite electrode. This gives a total electrostatic potential difference of $V = V_+ - V_-$ over the length L.

In conclusion, the simultaneous application of 0 to 30 kV to a reservoir and 0 to -30 kV to the opposite reservoir gives $0 \leq V \leq 60$ kV. As previously mentioned, considering that the ideal electric field strength should be in the $0 \leq E \leq 1000$ V cm^{-1} range. Therefore, the ideal distance between opposite microholes should be $L = 60$ kV$/1000$ V cm$^{-1} = 60$ cm. Consequently, an axial torus length ($2L$) of approximately 120 cm long is ideal.

The procedures that can be used to distribute the high voltages among the four electrodes will be discussed in the following subsections.

6.3.2 High Voltage Distribution Made by Four Modules

Four independent high voltage modules, each containing remote voltage regulation, remote current regulation, and each connected to a reservoir, can be used to make the toroidal system work. Figure 6.1 illustrates this system of four reservoirs, four electrodes, and four high voltage modules. If a 0 V dc signal is applied to the voltage regulation pin then it is ready for current regulation by applying 0 to 5 V dc to the current regulation pin (giving 0 to 100 % rated current output). Conversely, if a 0 V dc signal is applied to the current regulation pin then it is ready for voltage regulation by applying 0 to 5 V dc to the voltage regulation pin (giving 0 to 100 % rated voltage output).

In the electrokinetic active mode of operation this setup only works when high voltage modules with the option of voltage regulation and current regulation are used. They can be quickly remotely switched from voltage control to current control, and vice versa. In this case, a small current can be injected into the intermediate reservoirs (using remote current control) and the other reservoirs are voltage controlled. With this setup everything can be controlled all the time without the need for any moving parts. However, these high voltage modules with voltage

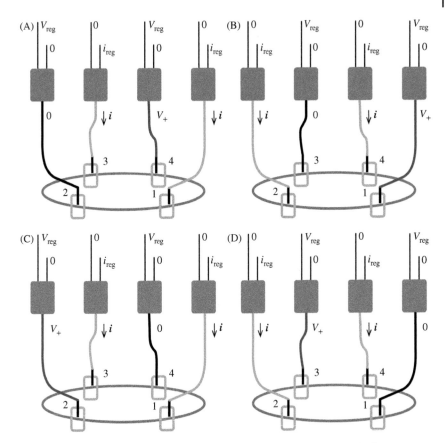

Figure 6.1 Schematic view of four high voltage modules connected to the four reservoirs in an EST with a toroidal layout. The sequence of dc signals that must be applied to the regulation pins of the high voltage modules in order to drive the analytes in a *quasi*-continuous circulating mode along a toroidal separation path are shown.

regulation and current regulation are more expensive. Even worse, four of these modules are required in this setup, making this option very expensive.

When the gravimetric or hydrodynamic active mode of operation is used then the above described option is no longer good. In this case the high voltage cables must be disconnected from the electrodes that high voltages are not applied to (intermediate reservoirs), otherwise they will not float because a residual electrostatic potential of a few hundred volts (in the regulated voltage mode), or a residual current (in the regulated current mode), will always be present. Moreover, these four bipolar high voltage modules make the system very expensive. The use of two voltage regulated high voltage modules are a much more convenient and cheaper solution, which is presented in the next subsections.

6.3.3 High Voltage Distribution Based on Relays

The use of two unipolar high voltage modules (one operating in the 0 to V_+ range and the other in the 0 to $V_- = -V_+$ range) and four high voltage relays is very convenient. This system can be used as a high voltage distributor which distributes the voltages among the correct electrodes at the right times. Both

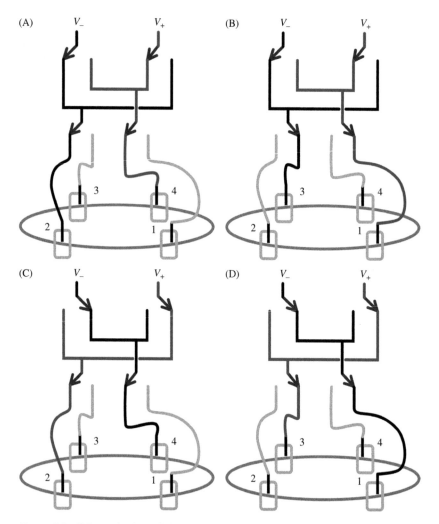

Figure 6.2 Schematic view of the switching sequence that must be used when four relays and two high voltage modules are used in an EST with a toroidal layout. The sequence of relay positions that must be used in order to drive the analytes in a *quasi*-continuous circulating mode along a toroidal separation path are shown.

mechanical (switches) and solid state relays can be remotely controlled and made to distribute the desired high voltages to the electrodes in the reservoirs in a timely and precise manner. Such a relay system was used in the pioneering work of Jorgenson et al. [12,13] Figure 6.2 shows their wiring and the relays they used. Attention should be paid to the quality of the relays: they must comply with both the voltage operation range (for instance 0 to 30 kV) and the low current range (otherwise they will be too large and expensive). In addition they must be totally isolated when they are switched off (no current leaking). The great majority of the commercially available relays do not meet these three prerequisites. Moreover, caution should be taken to prevent short circuiting at the relay outputs.

6.3.4 High Voltage Distribution Based on Sliding Switches

Figure 6.3 illustrates a linear version of the high voltage distributor. This is simpler to make and use; however after reaching the extent of its range of movement it must be quickly moved back across all of its previous positions to reach its starting position again. This should not affect functionality as the displacement can be completed in a fraction of a second. Moreover, all of these movements can be automated using a servo-motor and a ballscrew. The megaOhm resistors are used to inject a previously defined current. This is the procedure used in the electrokinetic active mode of operation (explained in Section 4.7.3). These megaOhm resistors must be removed and the circuitry must be left open (with the tip of the cables capped) if the gravimetric or hydrodynamic active modes of operation are to be used. A hand-driven distributor is the cheapest to construct. Moreover, the

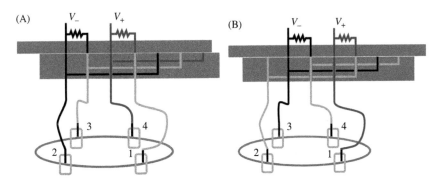

Figure 6.3 Schematic view of the positions that must be used to energize the four electrodes of an EST with a toroidal layout. The sequence A → B → C → D → A → ... must be used to drive the analytes in a *quasi*-continuous circulating mode along a toroidal separation path. This high voltage distributor is compatible with all of the active modes of operation (used to prevent band leaking): gravimetric, hydrodynamic, and electrokinetic.

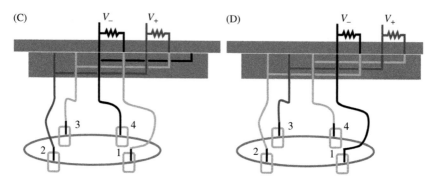

Figure 6.3 (*Continued*)

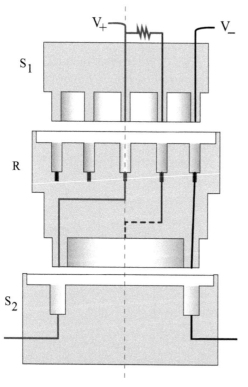

Figure 6.4 Illustration of the vertical cut of a rotating high voltage distributor which drives analytes in a *quasi*-continuous circulating mode by rotating the central rotor R (always in the same, direction) at specific times. This high voltage distributor is compatible with all of the active modes of operation (used to prevent band leaking) and is easy to automate.

automation of this system requires a servo-motor, a ballscrew, and a software program to control it.

6.3.5 High Voltage Distribution Based on Rotating Switches

A rotating version of the high voltage distributor is shown in Figures 6.4 and 6.5. It consists of three polymer pieces of cylindrical symmetry, closely fitted together: a static top (S_1), a static base (S_2) and a rotating center piece (R). Rotation of R by 90° and maintaining the same sense of rotation allows for switching of the high voltages conveniently to the four reservoirs of the toroid in a sequential fashion. The three building blocks of this connector are properly shaped with grooves of rectangular cross section and opposite shaped rims filling the voids, as shown

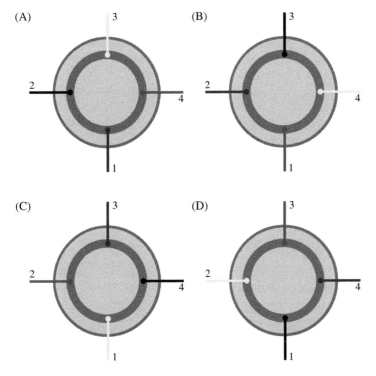

Figure 6.5 Top view of the static base S_2 of Figure 6.4. This figure shows the ligation used when cations are being analyzed, with sample injection into the microhole of reservoir 1, and the runs performed in a clockwise manner. The voltage distribution produced by each one of the four positions of the rotor R are shown. These outputs are directed to reservoirs 1 to 4. The rotor is rotated in a clockwise manner and at regular time intervals to produce the outputs given by positions A → B → C → D → A → ... and so on.

in Figure 6.4, allowing for good high voltage isolation, thus preventing any high voltage flash over. The properly configured internal electrical connections of R allow for the proper high voltage re-distribution from the upper sliding connection to the lower discrete point contact connection of the reservoirs of the toroid, as shown in figure 6.5. This concept of construction makes it easy to automate.

Figure 6.5 shows a top view of the polymer made (e.g., PTFE) static base cup. The high relief of the bottom of the rotor enters in this circular well and distributes the voltages (or current) according to the positions shown. The high voltages are redistributed at each quick rotation of 90°.

References

1 Blanes, L., Coltro, W.K.T., Saito, R.M. et al. (2012). *Electrophoresis* 33: 893–898.

2 Blanes, L., Coltro, W.K.T., Saito, R.M. et al. (2013). Practical considerations for the design and implementation of high-voltage power supplies for capillary and microchip capillary electrophoresis. In: *Capillary Electrophoresis and Microchip Capillary Electrophoresis: Principles, Applications, and Limitations* (ed. C.D. García, K.Y. Chumbimuni-Torres, and E. Carrilho). Hoboken, NJ: Wiley 67–75.

3 Hutterer, K.M., Birrell, H., Camilleri, P., and Jorgenson, J.W. (2000). *Journal of Chromatography B* 745: 365–372.

4 Hutterer, K.M. and Jorgenson, J.W. (2005). *Electrophoresis* 26: 2027–2033.

5 Henley, W.H. and Jorgenson, J.W. (2012). *Journal of Chromatography A* 1261: 171–178.

6 Henley, W.H., He, Y., Mellors, J.S. et al. (2017). *Journal of Chromatography A* 1523: 72–79.

7 Lee, S.J., Castro, E.R., Guijt, R.M. et al. (2017). *Journal of Chromatography A* 1517: 195–202.

8 Lopez-Canovas, L., Martinez Benitez, M.B., Herrera Isidron, J.A., and Flores Soto, E. (2019). *Analytical Biochemistry* 573: 17–29.

9 Goering, R.V. (2010). *Infection, Genetics and Evolution* 10: 866–875.

10 Maule, J. (1998). *Applied Biochemistry and Biotechnology – Part B Molecular Biotechnology* 9: 107–126.

11 Hjertén, S., Valtcheva, L., Elenbring, K., and Liao, J.-L. (1995). *Electrophoresis* 16: 584–594.

12 Zhao, J., Hooker, T., and Jorgenson, J.W. (1999). *Journal of Microcolumn Separations* 11: 431–437.

13 Zhao, J. and Jorgenson, J.W. (1999). *Journal of Microcolumn Separations* 11: 439–449.

14 Kist, T.B.L. (2017). *Journal of Separation Science* 40: 4619–4627.

7

Heat Removal and Temperature Control

7.1 Introduction

Almost all ESTs are subjected to radial temperature gradients across the separation media due to the Joule effect. When a constant electrostatic potential difference is applied along a homogeneous medium, and when the ionic current reaches a constant value, a constant heat is generated per unit volume and unit time (p). In this case, a constant temperature difference (ΔT) between the hottest line (usually the center-line) of the separation medium and the platform inner wall surface will be observed. For cylindrical capillaries, with a inner radius of a, total length of L, and with a constant temperature T_a over all points of the inner surface, a temperature difference ΔT will be produced. This can be calculated by assuming $r = 0$ in equation 2.33. Therefore,

$$\Delta T = T_0 - T_a = \frac{pa^2}{4\lambda} = \frac{Vi}{4\pi L\lambda} = \frac{Ei}{4\pi\lambda}, \tag{7.1}$$

where λ is the heat conductivity of the separation medium (buffer solution, polymer solution, or gel, depending on the separation mode used), V is the electrostatic high voltage difference applied between the two inert electrodes, i is the observed electric (ionic) current, and E is the applied electric field strength. It is assumed that the high voltage drops by an amount of V along L where the electric field is constant. This equation shows that high thermal conductivities (like that of water) produce smaller ΔT while low electrical conductivities (e.g., that of organic solvents) lead to larger ΔT [1].

As shown by equation 7.1, ΔT does not change if V and L are increased in the same proportion. However, increasing the electric field strength E will increase ΔT. In addition, it is possible to show that the value of i usually increases with ΔT. This is the reason of the non-linearity observed in the i versus V plot above certain values of E. This plot is known as Ohm's law and this non-linearity is a signature of over-heating. Moreover, above certain values of ΔT an important extra band

Open and Toroidal Electrophoresis: Ultra-High Separation Efficiencies in Capillaries, Microchips, and Slabs, First Edition. Tarso B. Ledur Kist.
© 2020 John Wiley & Sons Ltd. Published 2020 by John Wiley & Sons Ltd.

dispersion mechanism will appear and start to degrade the separation efficiency. This is called thermal peak dispersion and it is caused by the Joule effect that produces the radial temperature gradient. This happens mainly because the electrophoretic velocity of the analytes will be significantly higher at the center-line of the separation medium (the hottest location, with the lowest viscosity) than closer to the wall (the coldest location with higher viscosity) [2–4]. This dispersion coefficient due to the radial temperature gradients is denoted by D_{temp} in Section 2.13 and it is usually additive to molecular diffusion [5]: $d\sigma^2 = (2D + 2D_{temp})dt$. Therefore, there is a limiting value of the applied electric field strength that can be applied because above this value the performance indicators shown in Tables 5.4 and 5.5 start to degrade due to this thermal dispersion. This is unfortunate, as many important performance indicators depend on E and even on $E^{3/2}$.

Decreasing the cross sectional area of the separation medium is the traditional method of choice that allows the applied electric field strength to be increased to some extent, without causing too much band dispersion. Halving a will decrease ΔT by a factor of four, as shown by equation 7.1. However, this causes some handling problems as the separation media (with an ID of 10 μm or less) can became clogged. Moreover, fluorescent-labeled analytes are usually required in this case to overcome the poor detection limits resulting from the use of small sample volumes in such tiny conduits.

Finding other ways to mitigate the band dispersion caused by ΔT will most likely allow the application of stronger electric fields, which in turn will increase many performance indicators of the ESTs. This is still an unexploited field and the following subsections, while presenting the fundamental concepts and phenomena related to heat removal and temperature control, aim to stimulate those new possibilities.

Equation 7.1 assumes that p is constant and, consequently, the current density (j) is constant and independent of r ($0 \leq r \leq a$). However, the current density will be higher at the hottest points (usually at the center-line) and lower at the cooler points (close to the wall). This is true for the majority of the separation media used in the ESTs. An unexplored exception of that would be, for instance, if the pK_a of the BGE constituents are sensitive to temperature and, in addition, if the degree of ionization of the BGE constituents are lower at higher temperature (hottest points) compared with lower temperature (cooler points). This strategy will prevent ΔT from increasing too much, allowing higher electric field strengths to be applied. Another unexplored exception related to the previously cited, is the use of buffer solutions with a pH that is sensitive to temperature. The right choice of buffer and pH for a given class of analytes allows the mobility of the analytes to be modulated, so that they will be lower (smaller effective mobility) at the hottest points and higher at the coolest points. This strategy does not decrease ΔT, but it mitigates the band dispersion of the analytes caused by ΔT.

It must be remembered that heat is continuously generated in all ESTs. This heat diffuses out from the separation medium, through the platform wall and into the circulating coolant (gas or liquid). After a few seconds of operation, the temperature profiles reach a stationary state. Temperature gradients occur as a side effect of all ESTs and the only measure that has been taken to control them, both in normal commercial instruments and in prototypes, is to remove the heat from the system in an improvised manner. It is very rare to find a heat removal setup in the literature that produces, for instance, centrosymmetric temperature profiles.

The content of all previous chapters has been presented while paying special attention to the performance of the ESTs, i.e. by highlighting the operational parameters that can be used to optimize the performance indicators (summarized in Tables 5.4 and 5.5). This structure will be maintained in the present and the following chapters. At the beginning of this chapter, the commonly used heat removal setups are briefly presented and their resulting radial and axial temperature gradients are discussed. Finally, it is mathematically proven that the rational design of the radial temperature profiles across the separation media can be created by rationally designing the cooling geometries. These predictable and well defined temperature profiles have important implications in the performance of the ESTs and they present many new and unexplored possibilities.

Heat removal strategies are highly dependent on the platform used, but they do not depend on the layout (open or toroidal) or the separation mode used (see Appendix A), except for some technological details (two ends in the open layout and a continuous track in the toroidal layouts). Therefore, the content of this chapter will be discussed in the context of the three platforms: capillary, microchip, and slab.

7.2 Temperature Gradients are Unavoidable

The injection of a sample containing charged analytes requires a conductive separation medium, i.e. a liquid or gel containing dissolved cations and anions (the so-called BGE). This separation medium cannot be made of an organic non-conductive solvent because the injected sample plug would be subjected to an anti-stacking effect and disperse, as shown in Subsection 2.14. This happens because the electric field will be much weaker inside the injected sample plug than in the flanked non-conductive or low conductive organic solvent. The result is that the band will be subjected to a strong dispersion exhibiting a right angle triangular shape, with a sharp edge on one side and a long tail on the opposite side. This field distortion is caused by the enormous conductivity difference between the injected sample plug and the non-conductive organic solvent. In fact symmetric, narrow, and well behaved bands are only possible when the following two conditions are met: (1) When the conductivity of the sample is similar to or lower than that of the buffer. (2) When at least one BGE cation has an electrical mobility that is similar to the mobility of the cations under analysis, or when at least one BGE anion has an electrical mobility that is similar to the mobility of the anions under analysis. The isotachophoresis separation mode is an exception as it operates under the opposite conditions: (a) When cations are analyzed, the leading electrolyte cation has a mobility that is higher than the mobility of the fastest cation of the sample. The terminating electrolyte cation has a lower mobility than the slowest cation of the sample. (b) When anions are analyzed, the leading electrolyte anion has a mobility that is higher than the mobility of the fastest anion of the sample. The terminating electrolyte anion has a lower mobility than the slowest anion of the sample. This creates an enormous electric field distortion leading to the unique separation of the sample components as a train of zones with an increasing order of mobility (see Section 2.15.6).

Samples made of neutral and non-charged compounds can only be analyzed if at least one component of the BGE is charged. This charged component may be a micellar pseudo-phase, a microemulsion pseudo-phase, or any charged additive that interacts with the samples' neutral components (e.g., a charged inclusion complexion agent). Without a charged pseudo-phase the neutral components of the injected sample will not move and no separation will occur.

In the case of electrochromatography of neutral compounds, where electroosmosis is the only electrokinetic phenomena used, there must be a charged wall (electrical double layer) for the EOF to be present. Therefore, at least for this, an electric current will be observed (this surface current is the opposite of the electrokinetic phenomena called streaming current). But note that usually a buffer solution must be used to fix the pH, and this, in turn, increases the current and heat generated by a large amount.

Therefore, because of at least one of the three reasons cited, the Joule effect will always be present in the ESTs. In conclusion, temperature gradients will always be observed during the separations in the ESTs. *There is no way out other than to live with and handle them properly.*

7.3 Temperature has Multiple Effects

Temperature is regulated by controlling the temperature of the circulating coolant (usually between 0 °C and ~100 °C). Increasing the run temperature is well known for reducing the run time, as it reduces the dynamic viscosity of the liquids used to prepare the separation medium. This decrease of viscosity increases both the EOF and the mobility of the analytes.

Table 7.1 List of operational parameters that are affected by temperature and their observed impacts on separations given in their assumed decreasing order of importance

	Parameter	Impact
1	η	Decreases with temperature (increasing μ, μ_{eo}, and ionic current)
2	D	Increases with temperature and increases band broadening
3	pK_a	Affects the mobility of the analytes
4	pH	Affects the mobility of the analytes and EOF
5	K_{eq}	Effective mobility is impacted when using interacting additives
6	CMC	Changes with temperature
7	Solvation	Mean solvated H_2O usually decreases with temperature

There is always an optimal temperature, which gives a satisfactory separation in the shortest time interval, for a given set of operational conditions of an EST. However, it is hard to predict this optimal temperature when calculating from first principles as temperature has multiple effects on ESTs. Table 7.1 gives a list of operational parameters that are affected by temperature. They are listed in what is considered to be a decreasing order of importance.

Nevertheless, as a rule of thumb, lower temperatures generally favor the performance indicators in the following separation modes: AFE, ELFSE, FSE, IEF, ITP, MEKC, and MEEC. The main reason is that lower temperatures decrease the ionic current which, in turn, decreases the Joule effect and allows the application of stronger electric fields.

Sieving electrophoresis (SE) of oligonucleotides is an important exception as optimal resolutions are observed around 60 °C. For the separation of ssDNA (oligonucleotides and sequencing reaction fragments) and RNA, denaturing conditions are needed during the run to avoid strand hybridizations and the formation of secondary structures. This is obtained by addition of urea or formamide plus high temperature (60 °C or more). Lower temperatures have a second detrimental effect as they decrease the "reptation" migration regime which, in turn, decreases the mobility differences among oligonucleotides and leads to poorer resolutions.

The pK_a of some compounds may change by two units when the temperature of the solution is changed by $\sim$ 60 °C (for instance from 10 to 70 °C). The same happens with the pH of buffer solutions when they are prepared using some weak bases or weak acids (see Section 1.1.11). The pK_a changes that occur with temperature changes are usually expressed as dpK_a/dT. Table 7.2 gives the dpK_a/dT values of a list of compounds taken from Mandaji et al [6]. (which were compiled from many sources) that are commonly used to prepare buffer solutions to be

Table 7.2 List of acids and bases with their thermodynamic pK_a values, apparent (working) pK_a values, and dpK_a/dT values (temperature sensitivity) ranked in increasing order of dpK_a/dT.

	Compound	T (°C)	Therm. pK_a	App. pK_a	dpK_a/dT
1	1,3-Diaminopropane (pK_{a1})	25	8.64	8.78	−0.031
2	Amonia	25	9.25	9.2	−0.031
3	Piperidine	25	11.12	11.17	−0.031
4	Ethanolamine	25	9.5	9.45	−0.029
5	Tris	25	8.06	8.11	−0.028
6	N-Methyldiethanolamine	25	8.52	8.57	−0.028
7	Piperazine (pK_{a1})	25	9.73	9.87	−0.026
8	1,3-Diaminopropane (pK_{a2})	25	10.47	10.56	−0.026
9	Phosphate (pK_{a3})	25	12.33	12.1	−0.026
10	Diethanolamine	25	8.88	8.93	−0.025
11	Glycine (pK_{a2})	25	9.78	9.87	−0.025
12	D(-)-N-methylglucamine	20	9.52	9.57	−0.023
13	Tricine	25	8.15	8.1	−0.021
16	Bis-Tris	25	6.46	6.51	−0.02
17	ACES	25	6.9	6.85	−0.02
18	TES	20	7.5	7.45	−0.02
19	DIPSO	25	7.6	7.55	−0.02
20	TAPSO	–	7.71	7.66	−0.02
21	Triethanolamine	25	7.76	7.81	−0.02
22	TAPS	–	8.41	8.36	−0.02
23	Bicine	20	8.35	8.3	−0.018
24	CHES	20	9.3	9.25	−0.018
25	CAPSO	–	9.71	9.66	−0.018
26	CAPS	–	10.4	10.35	−0.018
27	BES	20	7.15	7.1	−0.016
28	Piperazine (pK_{a2})	25	5.68	5.82	−0.015
29	N-Methylpiperazine	25	4.75	4.7	−0.015
30	MOPSO	25	6.95	6.9	−0.015
31	HEPES	20	7.55	7.5	−0.014

Table 7.2 (Continued)

	Compound	T (°C)	Therm. pK$_a$	App. pK$_a$	dpK$_a$/dT
32	HEPPSO	–	7.8	7.75	−0.014
33	MES	25	6.15	6.1	−0.011
34	ADA	20	6.6	6.46	−0.011
35	PIPES	20	6.8	6.66	−0.009
36	Carbonic acid (pK$_{a2}$)	25	10.33	10.05	−0.009
37	Boric acid (pK$_{a1}$)	20	9.14	9.09	−0.008
38	MOPS	20	7.2	7.15	−0.006
39	Carbonate (pK$_{a1}$)	25	6.37	6.3	−0.0055
40	Phosphate (pK$_{a2}$)	25	7.21	6.93	−0.0028
41	Citrate (pK$_{a1}$)	25	3.13	3.08	−0.0024
42	Glycine (pK$_{a1}$)	25	2.35	2.29	−0.002
43	Butanedioic acid	25	4.21	4.16	−0.0018
44	Formic acid	25	3.75	3.7	0.000
45	Acetic acid	25	4.75	4.7	0.000
46	Chloroacetic acid	–	2.85	2.79	+0.0023
47	Phosphoric acid (pK$_{a1}$)	25	2.15	2.09	+0.0044

used in the ESTs. However, these dpK$_a$/dT values are not constant over very broad temperature ranges because they experience small changes below and above room temperature. Therefore, the ideal way of presenting Table 7.2 data is as an array of columns, with each column showing the dpK$_a$/dT of the compound for a given temperature (for instance at 5 °C intervals). Alternatively, these dpK$_a$/dT values can be expressed (at least in the 0 to 100 °C temperature range) according to the following equation:

$$\frac{dpK_a}{dT} = \frac{A}{T^2} + \frac{B}{T} + C, \tag{7.2}$$

where A, B, and C are constants and T is the absolute temperature (in Kelvin). This fits the data well and it originates from the following two facts:

(1) Firstly, equilibrium constants (e.g., K$_a$) vary with temperature according to the van't Hoff equation [7]:

$$\frac{d \ln K_a}{dT} = \frac{\Delta H^\ominus}{RT^2}, \tag{7.3}$$

where $\Delta H^\ominus$ is the standard enthalpy change of the reaction, R is the gas constant, T is the absolute temperature, and $\ln K_a = pK_a/\text{Log}(e)$. Note that the values of dpK_a/dT are negative for exothermic reactions (when standard enthalpy change is negative) and positive for endothermic reactions (when standard enthalpy change is positive).

(2) Secondly, the standard enthalpy change of a reaction is also a function of temperature. According to Kirchhoff's law of thermochemistry this is given by:

$$\left(\frac{\partial \Delta H^{\ominus}}{\partial T}\right)_p = \Delta C_p, \tag{7.4}$$

where ΔC_p is the heat capacity change under a constant pressure (the heat capacity difference between products and reactants). Taking equations 7.3 and 7.4, it is possible to show that equation 7.2 gives the curve of best fit for the dependence of dpK_a/dT on temperature (T). In this case, a table with four columns (containing compound name and constants, A, B, and C) will allow the determination of the values of dpK_a/dT in the 0 to ~100 °C range.

The list of acronyms used in Table 7.2 are shown in the footnote.[1]

Item 6 (dpK_a/dT) of Table 7.1 becomes very important if analytes exhibit a value of dpK_a/d$T \gg 0$ or a value of dpK_a/d$T \ll 0$. The same is true for item 4 (pH) of Table 7.1 if the buffer has a value of dpH/d$T \gg 0$ or a value of dpH/d$T \ll 0$. Note that the use of analytes and buffer solutions with opposite temperature sensitivities is one way to promote on-column band compression (see Section 2.14). The first few compounds and the last few compounds of Table 7.2 can be considered to have opposite temperature sensitivities. Therefore, they form ideal analyte–buffer or buffer–analyte pairs for on-column band compression procedures. This is also interesting for the toroidal layouts as it makes cyclic band compression possible (see Appendix G).

---

1 ACES, 2-(carbamoylmethylamino)ethanesulfonic acid; ADA, N-(2-acetamido)iminodiacetic acid; BES, (N,N-bis(2-hydroxyethyl)-2-aminoethanesulfonic acid); Bicine, 2-(Bis(2-hydroxyethyl)amino)acetic acid; Bis-tris, 2,2-Bis(hydroxymethyl)-2,2,2-nitrilotriethanol; CAPS, N-cyclohexyl-3-aminopropanesulfonic acid; CAPSO, 3-(Cyclohexylamino)-2-hydroxy-1-propanesulfonic acid; CHES, N-Cyclohexyl-2-aminoethanesulfonic acid; DIPSO, 3-(N,N-Bis[2-hydroxyethyl]amino)-2-hydroxypropanesulfonic acid; HEPES, (4-(2-hydroxyethyl)-1-piperazineethanesulfonic acid); HEPPSO, N-(Hydroxyethyl)piperazine-N'-2-hydroxypropanesulfonic acid; MES, 2-(N-morpholino)ethanesulfonic acid; MOPS, 3-(N-morpholino)propanesulfonic acid; MOPSO, 3-morpholino-2-hydroxypropanesulfonic acid; PIPES, piperazine-N,N-bis(2-ethanesulfonic acid); TAPS, N-Tris(hydroxymethyl)methyl-3-aminopropanesulfonic acid; TAPSO, 3-[N-Tris(hydroxymethyl)methylamino]-2-hydroxypropanesulfonic acid; TES, 2-[[1,3-dihydroxy-2-(hydroxymethyl)propan-2-yl]amino]ethanesulfonic acid; Tricine, N-(2-Hydroxy-1,1-bis(hydroxymethyl)ethyl)glycine.

7.4 Electrical Insulators with High Thermal Conductivity

Solid dielectric materials are used in the fabrication of EST platforms. Fused silica is the most commonly used material in capillary platforms, whereas borosilicate glass and polymers are commonly used in microchip platforms, and poly(methyl methacrylate) is used in the chambers of slab platforms. Poly(methyl methacrylate) is commonly known as Plexiglass®, which is its most well-known trade name.

The ideal platform material must exhibit a high thermal conductivity, high electrical resistivity (be a good electrical insulator), mechanical resistance, and chemical inertness (it should not react with water and the additive or organic modifiers used in the buffer solutions). In addition it must be easy to mold (can be extruded into microtubes and slices). However, it is difficult to find materials with all of these properties, especially with both high thermal conductivity and high electrical resistivity, as these properties are usually negatively correlated.

Platforms made of dielectric solids with high thermal conductivities are highly desired as they minimize temperature difference between the inner wall (separation medium/platform wall interface) and the external wall over which the coolant circulates (platform wall/coolant fluid interface). Figure 7.1 shows as a histogram, the thermal conductivity coefficients of some dielectric materials which have a thermal conductivity coefficient that is higher than fused silica [8,9]. Figure 7.2 compares the thermal conductivity of important dielectric polymeric materials with fused silica [8,9]. The highest thermal conductivity coefficients are observed for BeO, AlN, BN, Si_3N_4, SiC, and Al_2O_3. However, it is difficult to make capillaries from these materials for many reasons, such as, some of their melting points are too high, some of them are not chemically inert as fused silica, most of them have a breakdown voltage that is much lower than fused silica, and the dust of BeO is very toxic.

Dielectric fluids (gas and liquids) are used as circulating coolants in ESTs. Many systems use stagnant air, some use forced air, and a few use a recirculating liquid that is temperature controlled by a Peltier system or a thermal bath. Those liquid coolants must exhibit a few important properties that are listed in Table 7.3. It is hard to find a liquid chemical compound or even a mixture with all of these properties. Fluorinated alcanes or fluorocarbon based oligomers are commonly used as they exhibit some of the cited properties. Nevertheless, the search for dielectric liquids with a low viscosity and high thermal conductivity is an active field of research in material science.

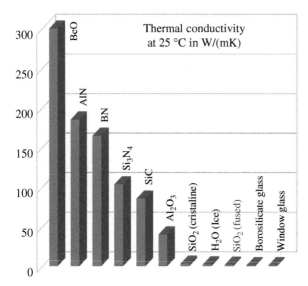

Figure 7.1 Histogram showing the thermal conductivity of fused silica (blue) compared to ice, glasses, and some solid dielectrics that exhibit very high thermal conductivities.

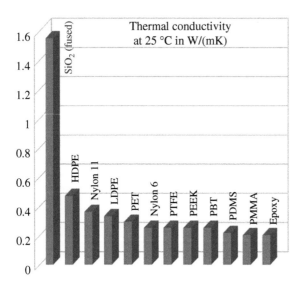

Figure 7.2 Histogram showing the thermal conductivity of fused silica (blue) compared to some dielectric thermoplastic and thermorigid polymers that are commonly used in the microchip and slab platforms. High density poly(ethylene), HDPE; low density poly(ethylene), LDPE; poly(ethylene terephthalate), PET; poly(tetrafluoroethylene), PTFE; poly(ether ether ketone), PEEK; poly(butylene terephthalate), PBT; poly(dimethylsiloxane), PDMS; and poly(methyl methacrylate), PMMA.

Table 7.3 List of properties that are important for the efficient performance of a recirculating liquid coolant in ESTs.

Property	Reason
Low dynamic viscosity	To allow high volumetric flow rates
High specific heat	High capacity to carry thermal energy away
High thermal conductivity	Facilitates heat diffusion into the coolant
Low flash point	For security reasons (non-flammables are ideal)
High electric resistivity	Insulators can flow very close to separation medium
Low melting point	Must be lower than that of water
High boiling point	Must be higher than that of water

7.5 Cooling Strategies Used in Capillary Electrophoresis

Figures 7.3 and 7.4 show the two most common cooling geometries used in commercial instruments. The forced air shown in Figure 7.3 creates an asymmetrical cooling system [10], which results in non-center-symmetrical isotherms inside the separation medium. The temperature profile that results from such cooling geometry is shown by the red asymmetric curve in Figure 2.19. A more interesting cooling strategy involves using a recirculating liquid coolant that flows along a polymer tube containing the capillary. This is illustrated by Figure 7.4. The drawback of this geometry is that the capillary fluctuates around the center-line of the polymer tube as it is difficult to keep it fixed in the middle. The resulting temperature profiles are not radially symmetric regarding the center-line of the separation medium because the circulating coolants do not flow symmetrically with respect to the capillary external wall. Moreover, the shear rate (see Section 7.7) of the coolant is low in this case, which means that there is a long distance between the points of highest speed of the coolant and the center-line of the capillary. As a result, a high temperature difference may be observed between the center-line of the capillary and the temperature set for the coolant, especially if the coolant has a low thermal conductivity. Finally, temperature is not controlled in either end of the open (common) capillary layout (injection inlet and detection outlet), which produces axial temperature variations along the capillary.

A close to ideal cooling geometry would lead to the following two results:

(1) The temperature difference between the recirculating coolant fluid and the surface of the inner wall (separation medium/platform wall interface) is small (5 °C or less).

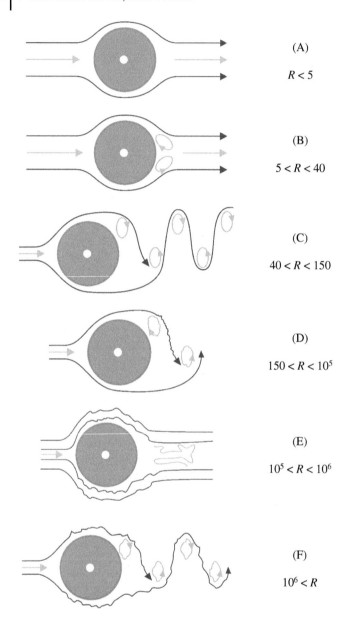

$$(A)$$

$$R < 5$$

$$(B)$$

$$5 < R < 40$$

$$(C)$$

$$40 < R < 150$$

$$(D)$$

$$150 < R < 10^5$$

$$(E)$$

$$10^5 < R < 10^6$$

$$(F)$$

$$10^6 < R$$

Figure 7.3 Air in thermostated compartments flows orthogonally to the axial direction of the capillary. In normal instruments the Reynold's numbers (R) observed corresponds to (C) and (D), depending on the velocity of the forced air [10]. In this case temperature is not constant over the external surface of the capillary (surface/air interface). Such temperature variations cannot produce a constant temperature at all points of the inner interface (separation medium/capillary wall interface).

(2) The resulting temperature at the surface of the inner wall (separation medium/platform wall interface) is the same and consistent at all points of this interface.

The advantages of condition 1 are two-fold: first, the runs can be performed at low temperatures (e.g. at 5 °C) by using a coolant at ~0 °C. Most buffer solutions will freeze between runs when a coolant with a temperature of below 0 °C is used. Second, stronger field strengths can be applied without the risk of the temperatures at the center-line reaching the boiling point of the solution used in the separation medium, which would interrupt the run. It will be shown later that this is important when Poiseuille counter-flow, which can cope with high temperature differences, is used.

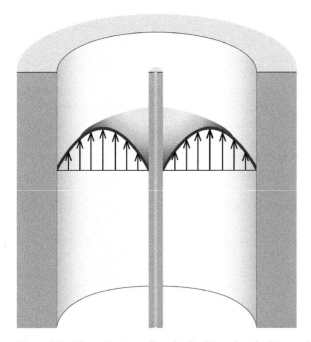

Figure 7.4 The velocity profile of a liquid coolant inside a polymer tube when the capillary is positioned in the middle of this tube. This cooling set-up is much better than the forced air shown in Figure 7.3. The dimensions are in scale for a capillary with an OD of 365 μm and a coolant tube with an ID of 5 mm. In this case a large temperature difference is still observed between what the coolant is set to and what is observed at the external surface of the capillary. Moreover, it is difficult to keep the capillary in the center of the tube. In practice it is randomly positioned anywhere inside the polymer tube. This turns very difficult to produce a constant temperature on all points at the separation medium/inner wall interface.

The advantage of condition 2 is that it allows the rational design of new separation methods in which the dependence of the buffer pH on temperature is used, for instance, to slow the mobility of the analytes at the center-line (the hottest place with the lowest viscosity). This way, the band dispersion caused by ΔT can be diminished or mitigated. The same strategy can be used when the pK_a of the analytes is sensitive to temperature while the pH of the buffer is not. Finally, the Poiseuille counter-flow strategy can also be used to mitigate band broadening by applying a counter-pressure or gravimetric driven flow along the capillary (described in Appendix F).

These rational cooling geometries are the subject of all of the following sections and subsections. They are presented separately for the capillary, microchip, and slab platforms.

7.5.1 Advantages of a Symmetric Cooling Geometry

An important advantage of using a symmetric rational cooling strategy is that the separation method can be designed to mitigate band broadening and consequently increase the separation efficiency. Buffers with a temperature sensitive pH can be used to decrease the mobility of the analytes at the hottest points of the separation paths, which are usually found along their center-line. Consequently, a given analyte will have the same velocity at each point across the separation path, regardless of the presence or absence of the temperature gradients caused by the Joule effect. Therefore, much stronger electric fields can be applied, which will improve the performance indicators (shown in Tables 5.4 and 5.5).

Another important advantage of such symmetric cooling systems is that the opposite of what has just been described can occur; the separation method can be designed to mitigate band broadening by using the temperature sensitivity of the analytes' pK_a to increase separation efficiency. This technique can be used to reduce a given analytes' mobility along the center-line of a separation path. Therefore, the molecules of a given analyte will have the same velocity at each point across the separation path, regardless of the presence or absence of the temperature gradients caused by the Joule effect. Again, much stronger electric fields or higher voltages can be applied, and consequently performance indicators (shown in Tables 5.4 and 5.5) will be improved.

In addition to the previous general advantages of rationally designed buffers, rational symmetric cooling systems have a third impressive advantage. They allow the band broadening that results from high radial temperature gradients (ΔT) to be mitigated using a Poiseuille counter-flow. Here, a symmetrical cooling system is discussed based on the mathematical analyses of fluid flow (see also Appendix F). From the theoretical point of view it has the potential to significantly increase separation efficiency, resolution, and peak capacity (narrower peaks). However, the cooling geometry must be designed rationally for this to work, as explained in the coming paragraphs (the mathematical deductions are left for Appendix F).

Figures 7.5 to 7.12 depict some examples of symmetric cooling geometries. The most important property of these cooling geometries is that they ensure an almost constant temperature at every point of the inner interface (liquid phase/fused silica wall). This border condition is essential for the Poiseuille counter-flow strategy to successfully mitigate the band broadening caused by temperature gradients (ΔT). In this case the pressure gradient or reservoir level difference can be calculated and they become new operation parameters. Simply put, the electrophoretic velocity difference (Δv_e) between the center-line and the wall, caused by the temperature difference (ΔT), will be canceled out by the velocity of the Poiseuille counter-flow, given by both equations 2.14 and 2.15. According to Appendix F this is true for the general case, i.e, for cylindrical, square, rectangular, or any other arbitrary continuous inner geometry. This means that, in modulus:

$$v_e(T_a + \Delta T) - v_e(T_a) = v_{max}. \tag{7.5}$$

This electrophoretic velocity (v_e) difference is given by equation 2.31. If the terms are arranged, it results in the following: $v_e(T_a + \Delta T) - v_e(T_a) = [\mu_e(T_a + \Delta T) - \mu_e$

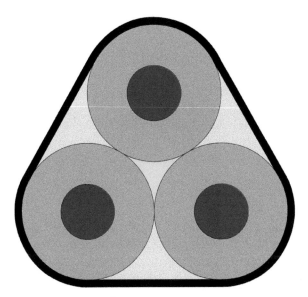

Figure 7.5 Example of a cooling geometry that produces an approximately constant temperature on the separation medium/wall interface. The blue circles represent the coolant, the light turquoise area represents the separation medium, and the black line represents a tape that holds the three capillaries together with the aid of a glue (yellow). All of the capillaries have an ID of 150 μm and an OD of 365 μm (model 150375). Note the proximity of the circulating coolant and the separation medium ($\approx 107 \ \mu$m).

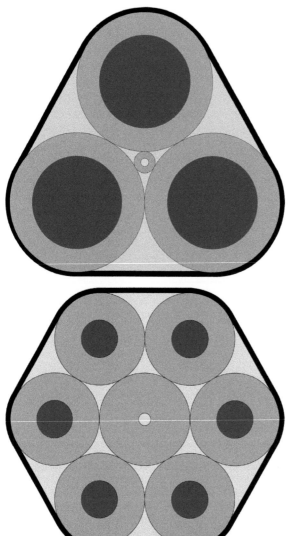

Figure 7.6 A symmetrical cooling geometry similar to Figure 7.5, but composed of four capillaries. The analytical capillary is thin and located in the middle, surrounded by three thick capillaries which hold the coolant (blue). In this illustration the central capillary has an ID of 40 µm and an OD of 105 µm (model 040105), while the coolant capillaries have an ID of 450 µm and an OD of 670 µm (model 450660).

Figure 7.7 Example of a symmetrical cooling geometry using seven capillaries. The central capillary has an ID of 50 µm and OD of 365 µm (model 050375) while the six external capillaries have an ID of 150 µm and an OD of 365 µ (model 150375). The blue circles represent the coolant, the light turquoise circle represents the separation medium, and the black line represents a tape that holds the capillaries together with the aid of glue (yellow). Note the proximity of the circulating coolant and the separation medium. Moreover, temperature is almost constant at the inner interface (separation medium/dielectric surface), which produces center-symmetrical cylindrical isotherm surfaces inside the separation medium.

Figure 7.8 A symmetrical cooling geometry consisting of a square capillary inside a cylindrical capillary. The blue area represents the coolant fluid and the light turquoise square represents the separation medium. The central capillary has an internal width of 50 µm and an external width of 365 µm (model 050375). The external capillary has an ID of 530 µm and an OD of 700 µm (model 530700).

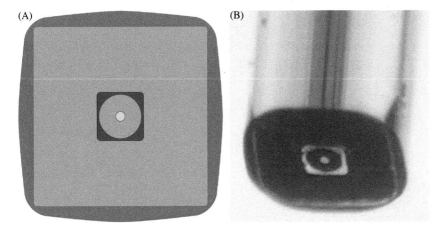

Figure 7.9 (A) Illustration of a symmetrical cooling geometry consisting of a cylindrical capillary inside a square capillary. This is the opposite of Figure 7.8. The blue area represents the coolant fluid and the light turquoise circle represents the separation medium. The central capillary has an internal diameter of 20 µm and an eternal diameter of 90 µm (model 020090). The external capillary has an internal width of 100 µm and an external with of 365 µm (model 100375). (B) Photograph of a cylindrical capillary inside a square capillary. The inner capillary is Molex/Polymicro's model TSP020090 and the outer capillary is model WWP100375. (Author's unpublished result.)

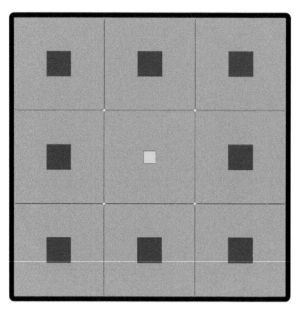

Figure 7.10 A symmetrical cooling geometry using square capillaries. The blue squares represent the coolant and the light turquoise square represents the separation medium. The central capillary has an internal width of 50 μm and an external width of 365 μm (model 050375). The surrounding capillaries have an internal width of 100 μm and an external with of 365 μm (model 100375).

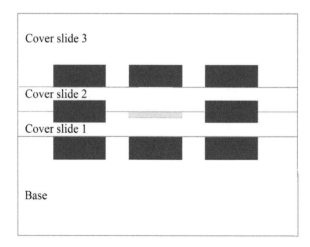

Figure 7.11 Microchip with an ideal cooling design. The coolant (blue) circulates through channels that are symmetrically positioned around the separation microchannel (light turquoise). Such a design produces a temperature that is consistent in all points of the separation medium/dielectric wall interface.

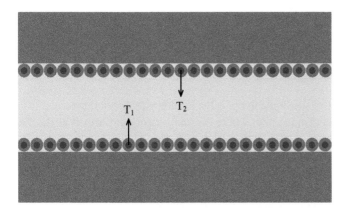

Figure 7.12 The wall of an ideal slab electrophoresis chamber. The separation medium is represented by the light turquoise. The coolant (blue) flows through the capillaries that are glued (yellow) against the electrophoresis chamber wall. In the case of a toroidal layout, the inner wall (shorter path) is maintained at a lower temperature (T_1) while the outer wall (longer path) is maintained at a higher temperature (T_2). Figures 4.5 and 4.6 illustrate the vertical and horizontal versions of the toroidal electrophoresis chambers.

$(T_a)]E = kE\Delta T\mu_e(T_a)$, which can be introduced into equation 7.5 to obtain:

$$\Delta P = \frac{k\eta V^2 i}{\lambda AL}\mu_e(T_a), \tag{7.6}$$

where ΔP is the pressure difference that must be applied between reservoirs that are connected by an arbitrary microconduit with length L and cross sectional area A, and among which an electrostatic potential difference V is applied. A liquid that has a viscosity of η fills the conduit with an arbitrary geometry, which can be cylindrical, square, rectangular, or any arbitrary shape, as illustrated by Figure F.1. Note that the conduit must have a constant cross sectional shape along its whole extension. For a detailed deduction see Appendix F. The parameter $\mu_e(T_a)$ denotes the mobility of the analyte at the positions with the lowest temperature, i.e. close to the surface of the separation medium/platform wall interface. The factor k expresses the mobility change in temperature, $k \simeq 0.025\ °\mathrm{C}^{-1}$.

An assumption was made in equation 7.6 where the viscosity of the liquid was considered constant when v_{max} was calculated and given by equations 2.14 and 2.15. However, when a radial temperature gradient is observed the viscosity will be a bit lower along the hottest line than at any point close to the wall, which makes equation 7.6 only an approximate result. Nevertheless, it is a very useful guide to find the optimal ΔP or ΔH in experiments.

An expression similar to equation 7.6 is obtained when a level difference of ΔH is used:

$$\Delta H = \frac{k\eta V^2 i}{\lambda \rho g A L}\mu_e(T_a),\tag{7.7}$$

where ρ is the density of the liquid and g is the local gravity acceleration.

Figure 7.5 shows an arrangement where the analyses are performed along the void between the three tightly bound cooling capillaries. Figure 7.6 shows a cooling geometry consisting of one thin capillary surrounded by three thick cooling capillaries. The same cooling set-up can be constructed using four, five, or even a generalized number of n cooling capillaries. However, in all cases all of the sequential outer capillaries need to be in contact with each other and each one needs to touch the inner capillary. This only happens when the ratio of the outer diameter of the thin capillary (od) and the outer diameter of the thick capillaries (OD) follow this relation:

$$\frac{OD}{od} = \frac{\cos(\alpha/2)}{1 - \cos(\alpha/2)} \quad \text{with} \quad \alpha = \pi\left(1 - \frac{2}{n}\right),\tag{7.8}$$

where n is the number of cooling capillaries used ($n = 3, 4, 5, \ldots$), OD is the outer diameter of the coolant capillaries, and od is the outer diameter of the thin, central analytical capillary. Following equation 7.8 a compact setup in which all neighbor capillaries touch each other is obtained.

Figure 7.6 shows an example where the minimum value of $n = 3$ cooling capillaries is used. Figure 7.7 shows an example where $n = 6$ is used. In this particular case of $n = 6$, according to equation 7.8, the other diameter of the cooling capillaries is the same as the analytic capillary (OD = od). Therefore, using commercially available capillaries with ID of 150 μm as coolants and ID of 50 or 75 μm as the analytical capillary is a good combination. The OD of all can be, for instance, 365 μm. It is important to use coolant capillaries with an ID and length that exhibits a good hydraulic resistance, i.e. not too high and not to low (see Section 4.5), this is important to help the liquid circulation and heat removal, otherwise a temperature difference between the inlet and the outlet of the coolant would be observed. If the ID is too large then the shear rate will be too low (see Section 7.7).

Inserting an analytical square capillary inside a cylindrical cooling capillary produces four gaps for coolant circulation, as shown in Figure 7.8. However, it is difficult in this case to assure a consistent temperature over all points of the separation medium/analytical capillary inner wall interface. Finally, inserting a cylindrical analytical capillary inside a square coolant capillary produces a similar effect, see Figure 7.9A. Figure 7.9B shows a photography taken from such a setup after polishing the tip at an angle of 90°.

7.6 Cooling Strategies Used in Microchip Electrophoresis

It is difficult to predict the temperature profiles in microchips that occur during the runs because this platform usually has a rectangular geometry and many channels, wells, and sometimes many other installed accessories. Therefore, in practical situations, the methods and protocols used in these systems are usually optimized by trial and error.

Many commercial instruments use a microchip with a thin base, and a flux of air is forced to hit this surface during the runs. The short distance between the separation microchannels and the circulating air makes temperature regulation relatively efficient. The temperature regulation can be even more precise if the air is recirculated within an oven with a regulated temperature (Peltier system). This can set the temperature to the optimal value for each method and application.

When the durations of the runs are short and thick walled microchip platforms are used (as usually happens when prototyping this platform) then the microchip's body works as a temperature reservoir. This is even more efficient if the chip has a high heat capacity, as only a small temperature increase may be observed during these short run times. In this case the inter-run time intervals are important because the chip's body must cool down to room temperature before the next run can begin.

7.6.1 Advantages of a Symmetric Cooling Geometry

The Poiseuille counter-flow technique (see Appendix F) can be successfully used in microchips if the cooling geometry assures a constant temperature over the separation medium/microchip inner wall interface. In this case, the level contour of the temperature isotherms (and consequently the level contour of the electrophoretic velocities) will coincide with the pressure driven isovelocity level contours. For rectangular conduits, which are the most common in microchips, this is illustrated by Figures 2.14, 2.15, and 2.16.

Figure 7.11 shows an example of a cooling geometry that produces an almost constant temperature over the separation medium/microchip wall interface. The blue areas represent the recirculating coolant and the light turquoise area represents the separation medium. This requires the fabrication and alignment of four layers that are then fused together.

7.7 Cooling Strategies Used in Slab Electrophoresis

The use of stagnant air is the standard cooling procedure used in the great majority of practical situations. This only works because of the natural convection of hot and cold air currents. However, both the performance indicators, repeatability and reproducibility of the separations, could benefit from the adoption of, at least, the forced air strategy. This is achieved by using a fan on each side of the slab. Temperature regulation, for instance at 20 °C, can be done by adjusting the air conditioning settings of the room where the runs are performed. Another strategy is to recirculate the buffer between the reservoirs and through a cooler, or to directly cool it by immersing cooling coils into the buffer tanks. However, this can be dangerous if not well isolated from the applied electrostatic potentials.

There are many other factors that affect the quality of the runs performed along slabs in chambers. One factor is the use of well leveled chambers during the runs, otherwise non-homogeneous electric fields will be produced along the slab, which would, in turn, distort the bands. The tray used to pour the gel must also be very well leveled during the pouring process. Samples may run out of the gel if the wedge of irregularly sized slabs of gel is in the bad direction.

7.7.1 Advantages of a Rational Cooling Strategy

An almost ideal way to control the temperature in slab electrophoresis is to cover the chamber's surface with an array of microtubes containing a coolant fluid which circulates through a thermo-regulated bath. This is especially advantageous in toroidal slab electrophoresis, where the difference in the lengths of the inner and outer paths can be compensated by applying a lower temperature along the inner path and a higher temperature along the outer path. A high level of band dispersion will occur if this is not done due to the bending of the slab. This is even more pronounced when thick walled slabs made of gels or other media are used. The same phenomena occurs in capillary electrophoresis when the capillaries are coiled [11].

7.8 Shear Rate of the Coolant

It is also necessary to use and apply the concept of shear rate in order to produce the rational and predictable temperature profiles described in Sections 7.5.1, 7.6.1, and 7.7.1.

If the space between two parallel plates, separated by a distance h, is filled with a fluid, and if one plate moves at velocity v and the other stays at rest, then the shear rate of the fluid is given by $\gamma = v/h$. Another simple example of shear rate is given by Figure 2.16. The shear rate of the fluid along the points of the vertical line that crosses the center ($y/h = 0$) is given by v_{max}/h.

A more general definition of shear rate is the following: it expresses the gradient of velocity inside a flowing material. In the case of a long conduit filled with a fluid that flows in the x direction, the shear rate is given by $\gamma = \partial v/\partial y + \partial v/\partial z$. For cylindrical capillaries it is possible to show that the shear rate is given by $\gamma = 4v/a$, where a is the capillary inner radius.

Cooling systems operated at low shear rates (cooling fluids with low v in capillaries with large a) tend to be problematic for many reason. First, a large temperature difference can be established between the central line of the cooling conduit and its internal surface. This creates large temperature differences between what is set on the coolant and what is actually observed at the separation medium/platform wall interface. This limits the working temperature range within the separation medium because the maximum temperature cannot exceed 100 °C when working with aqueous solutions. Second, it is difficult to assure a constant temperature over all points on the separation medium/platform wall interface when working with such large distances as small distance changes will affect the temperature profiles.

In conclusion, small cooling conduits that are placed close to the separation medium and filled with coolants that are operated at high shear rate will lead to the best results. The chemical inertness and optimal electrical insulating properties of fused silica makes this material optimal to be used as the coolant tubing. Moreover, cylindrical microtubes made by this material withstand high internal pressures, which is important in creating coolant fluid flow with high shear rate.

7.9 Final Considerations

Heat removal is mandatory in almost all ESTs in order to keep the repeatability of the quantitative analyses and performance indicators at their highest possible levels. It is unfortunate that heat removal has been performed in an improvised manner in almost all publications. This chapter, while presenting the subject of heat removal, has shown that making small adjustments to the coolant circulation geometry can lead to a consistent temperature over all points of the separation medium/platform wall interface. The many potentials of this consistent temperature was theoretically demonstrated for the situations where buffers with high temperature sensitive pHs are used or when analytes with high temperature sensitivity pK_a are being analyzed (Table 7.2). In another situation the advantages of using symmetric temperature profiles was theoretically demonstrated using models based on the Navier–Stockes equations and the heat diffusion equation. In this case, if the temperature is the same at all points of the separation medium/platform wall interface, the application of a Poiseuille counter-flow to the open flow capillary or microchannel mitigates the thermal band broadening caused by the Joule effect. This allows much stronger electric fields to be applied and the performance indicators 2a, 2b, 3a, 3b, 5a, 5b, 7a, 7b, 9a, and 9b of Table 5.4 can be improved all together. To the best of the present author's knowledge, this is the first time this is theoretically demonstrated in the literature for the general case (demonstrated in Appendix F). For cylindrical capillaries this was theoretically demonstrated by Gobie and Ivory [13]; moreover, they took into account the effect of temperature on conductivity and the respective "autothermal effect".

References

1 Porras, S.P., Marziali, E., Gaš, B., and Kenndler, E. (2003). *Electrophoresis* 24: 1553–1564.

2 Ghosal, S. (2004). *Electrophoresis* 25: 214–228.

3 Xuan, X. and Li, D. (2004). *Journal of Micromechanics and Microengineering* 14: 1171–1180.

4 Gaš, B., Štědrý, M., and Kenndler, E. (1997). *Electrophoresis* 18: 2123–2133.

5 Virtanen, R. (1993). *Electrophoresis* 14: 1266–1270.

6 Mandaji, M., Rübensam, G., Hoff, R.B. et al. (2009). *Electrophoresis* 30: 1501–1509.

7 Atkins, P.W. and de Paula, J. (2006). *Physical Chemistry*. Oxford University Press.

8 Huang, X., Jiang, P., and Tanaka, T. (2011). *IEEE Electrical Insulation Magazine* 27: 8–16.

9 Bird, R.B., Stewart, W.E., and Lightfoot, E.N. (2006). *Transport Phenomena*, revised 2nd edition. Hoboken, NJ: Wiley.

10 Lienhard, J.H. (1966). *Synopsis of Lift, Drag, and Vortex Frequency Data for Rigid Circular Cylinders*. Pullman, WA: Washington State University.

11 Kašička, V., Prusík, Z., Gaš, B., and Štědrý, M. (1995). *Electrophoresis* 16: 2034–2038.

12 Mandaji, M., Rübensam, G., Hoff, R.B. et al. (2009). *Electrophoresis* 30: 1510–1515.

13 Gobie, W.A. and Ivory, C.F. (1990). *Journal of Chromatography* 516: 191–210.

8

Detectors

8.1 Introduction

In this chapter the most widely used detectors of the open and toroidal layouts are presented and any adjustments that are necessary to make some of them compatible with the toroidal layout in the capillary, microchip, and slab platforms will also be explained.

Detection systems are separated into two groups: fixed point detectors, which give electropheorgrams (signals measured along time), and spatial detection systems, which reveal the spatial profile of the bands along the separation path at a given time during the run (measured with scanners and photographic cameras). The most used detection systems include absorption detection, contactless conductivity detection,[1] laser induced fluorescence detection,[2,3] and mass spectrometery.[4,5]

Derivatization reactions are normally used to improve detection limits, especially when trace amounts of analytes are analyzed using laser induced fluorescence. Moreover, this chapter shows that performance indicators can be significantly improved when some derivatization reactions are used not only in fluorescence detection but also in other detection systems including absorption detection and others.

Selectivity is the ability of a method to provide accurate results regardless of the presence of other substances (potential interferants).[6,7] A method with good separation selectivity produces clean chromatograms and electropherograms with high repeatability and reproducibility, regardless of the concentration of interferants. Therefore, a second goal of this chapter is to show ways of improving both detection selectivity and separation selectivity in ESTs when derivatization reactions are used, while describing the detection systems. It is difficult to quantify these quality criteria and for this reason they not treated in Chapter 5.

Open and Toroidal Electrophoresis: Ultra-High Separation Efficiencies in Capillaries, Microchips, and Slabs, First Edition. Tarso B. Ledur Kist.
© 2020 John Wiley & Sons Ltd. Published 2020 by John Wiley & Sons Ltd.

8.2 Fixed Point Detectors

Fixed point detectors are the simplest and most common detectors used. They are fixed at or close to the end of the separation path and do not require any moving parts.

In the open layout they are positioned as close as possible to the outlet in order to maximize the performance indicators, otherwise only a fraction of the applied high voltage will effectively contribute to the separations. This is represented by the term Vl/L in Table 5.4, where l/L is the fraction of the high voltage that is effectively used in the separations. The best case scenario is when $l = L$.

In the toroidal layout a detector can be positioned anywhere along the separation track. When two detectors are used they are positioned opposite each other. These two detectors may be identical or different. The advantage of using two detectors is that they maximize peak capacity, especially when analyzing samples with a large number of components with large mobility differences between a front running group and a slow moving group.

Absorption detectors are by far the most widely used systems. This is because they are cheap and behave almost as universal detectors. Nevertheless, they struggle to detect analytes with small molar absorption coefficients in microchannels and capillaries. Examples of such analytes include saturated fatty acids, saturated short chain carboxylic aicds, hydroxyl group(s) bound to aliphatic chains (alcohols), amine groups bound to aliphatic chains (primary amines), secondary amines made of aliphatic chains, and tertiary amines made of aliphatic chains, to mention just a few. Figure 8.1 shows the result produced by a fixed absortion detection system doted with a DAD detector. The spectra of the analytes are measured when they cross the detection point. In this example a mixture of adenosine (uncharged), adenosine monophosphate (AMP) (degraded), adenosine diphosphate (ADP), and adenosine triphosphate (ATP) was separated.

Absorption detectors are compatible with almost all layouts and platforms, as only a few require the insertion of the tip of the capillary into the detection holder. These particular detectors are not recommended for toroidal capillary electrophoresis because installing the capillary first and then joining the ends via fusion splicing is not practical.

Contactless conductivity detectors, at the time of writing, are gaining popularity and promise to dominate the market in the capillary and microchip platforms. Despite this, they have both pros and cons. Their main advantages are their contactless operation, low fabrication cost, low maintenance cost, robustness, and ease of operation. However their sensitivity is limited. A mobility difference between the passing bands and the BGE is required in order to produce good sensitivities. However, it is well known that good separations (symmetrical peaks) require BGEs to have constituents with similar mobilities to those of the

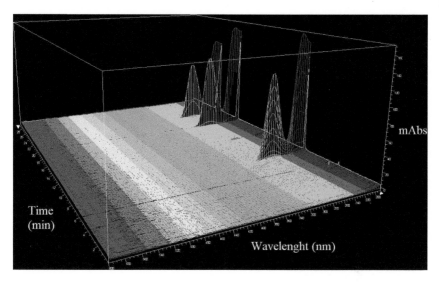

Figure 8.1 Electropherogram of a mixture of adenosine, AMP (degraded), ADP, and ATP using a DAD absorption detector that takes the absorption spectra of the analytes during their passage at the detection point. In this the absorption spectra was taken from 200 nm to 600 nm. The HP3D CE instrument used for this separation was equipped with a 75 µm ID capillary, 365 µm OD, 60 cm total length, and 25 mM phosphate buffer pH 7.0 as the separation medium (unpublished results from the author).

analytes. The following dilemma is normally observed: good separations with poor sensitivity or poor separations (asymmetric peaks) with good sensitivity.

Contactless conductivity detectors can, in general, be used in toroidal layouts. However, some models contain two silver O-ring shaped electrodes that are not compatible with the toroidal capillary layout. This is because the torus does not have an end that can be slid through these O-rings.

Fluorescence detectors are among the most sensitive detection systems. Moreover, the great variety of LEDs and diode lasers available make these detection systems very flexible, as the excitation light can either be guided by optical fibers or directed by lenses and mirrors. Until recently the predominant light detectors were photomultiplier tubes, but charged-coupled devices (CCDs) have become very sensitive and also very cheap. Avalanche photodiodes (APDs) also do a good job, however they are generally more expensive than CCDs.

The flexibility of fluorescence detectors makes them compatible with all platforms and layouts. However, in general they are more expensive than absorption detectors and C^4D, as they require many more optical parts including an excitation source, an interference filter at the outlet of the laser or LED, lenses to focus the excitation radiation onto the separation track, lenses to harvest the fluorescent light, longpass filters to block the scattered excitation radiation from getting into the detector, and finally a light detector.

Mass spectrometry works with mass-to-charge ratio of ions and is compatible with the open layout. This technique revolutionized the studies of proteomes,[8] metabolomes, food analyses, pharmaceutical analyses, and many other fields.[4,5] In the case of toroidal capillary electrophoresis, an important adaption is required to make this detection system possible: a piece of guiding capillary with a conical tip that connects a microhole to the inlet of the mass spectrometer must be installed. Moreover, the active mode of operation, which is used to prevent the bands from leaking out of the torus, must be paused when the bands are driven from the inner lumen of the torus to the mass spectrometer. In addition, electroosmosis must be used in the transference capillary (by applying a proper high voltage at the tip of the capillary where the electrospray is generated) in order to quickly transfer the bands for analyses from the torus to the mass spectrometer. This procedure turns, in principle, many mass-spec based hyphenation techniques compatible with the toroidal layout. Examples of such resulting hyphenated techniques would be: TCE-ESI-MS, TCE-ICP-MS, TCE-MALDI-MS, and sheath flow and sheathless interfaces. These potential hyphenations would join the benefits of toroidal electrophoresis (high peak capacity and resolving power) with mass spectrometry (high sensitivity and it effectively characterizes the components with high certainty).

8.3 Spatial Detectors (Scanners and Cameras)

Spatial detectors are triggered at defined instants of time and take a snapshot of the band distributions along the separation path. These detectors are divided into two groups: the camera-like detectors (with no moving parts) and the scanner-like detectors (that move along the separation path).[9–12]

Camera-like detectors are very common in the slab platform. They use a transilluminator, which either increases the contrast between the bands and their surroundings (when staining techniques are used) or excites the fluorophores (when fluoresence detection is used).

Camera-like detectors are compatible with the open and toroidal layouts, and are easy to use in slab and microchip platforms. However, it is difficult to use them with long capillaries (50 cm or more) since the camera must be moved away from the capillary in order to capture the whole image. This reduces the solid angle of the harvested fluorescent light, which results in poor detection limits. In addition, the excitation of analytes within such long capillaries is challenging.

8.4 Derivatization Reactions

Performing pre-column derivatization reactions with the analytes (before their separation) has been considered a demanding task that increases costs and may cause errors in quantitative analyses. However, these procedures have experienced great development over the years, resulting in reliable and convenient protocols.[13–15] The advent of automated dispensers and the possibility to keep the derivatization solutions permanently stored in inert, refrigerated bags, brings many advantages.

Both the electrical mobility (μ_{el}) and the diffusion coefficient (D) usually change when the derivatization reactions are performed. The diffusion coefficient always decreases while the electrical mobility may change in any direction and may even be subjected to a sign reversal, depending on the analyte, derivatizing reagent, and pH of the separation medium used. The performance indicators depend on these operational parameters, as shown in Table 8.1.

Table 8.1 Dependence of the performance indicators on the apparent mobility (μ) and molecular diffusion coefficient (D). The apparent mobility is defined as $\mu = \mu_{el} + \mu_{eo}$.

Performance indicator	Symbol	Dependence
Number of plates	N	$\propto \dfrac{\mu}{D}$
Plate double-rate	$\mathcal{N}$	$\propto \dfrac{\mu^3}{D}$
Plate height	H	$\propto \dfrac{D}{\mu}$
Resolution	R	$\propto \Delta\mu \left(\dfrac{1}{\overline{\mu D}}\right)^{1/2}$
Resolution rate	R/t	$\propto \Delta\mu \left(\dfrac{\overline{\mu}}{D}\right)^{1/2}$
Band capacity	B	$\propto \left(\dfrac{\mu}{D}\right)^{1/2}$
Band rate	B/t	$\propto \left(\dfrac{\mu^3}{D}\right)^{1/2}$
Peak capacity	P	$\propto \left(\dfrac{\mu}{D}\right)^{1/2}$
Peak rate	P/t	$\propto \left(\dfrac{\mu^3}{D}\right)^{1/2}$

The electroosmotic flow observed in fused silica capillaries and in microchips fabricated out of borosilicate glass runs in the direction of the cathode (it flows from the positive electrode to the negative electrode – from the higher electric potential to the lower electric potential). Therefore, positively charged derivatives perform much better than the negatively charged because their electrical mobility (μ_{el}) will point in the same direction of electroosmotic mobility, resulting in high apparent mobilities. The only exception is resolution, which is proportional to $1/\sqrt{\mu}$, as shown in Table 8.1.

The relationships between the performance indicators and both μ and D are expected to be valid for all layouts, all platforms, and many separation modes (ELFSE, FSE, AE, and, to some extent, both MEKC and MEEKC). Therefore, positively charged derivatives with high apparent mobilities are expected to perform well in all layouts, platforms, and most of the separation modes when high band capacity, high peak capacity, and resolution per unit time are desired (see Table 5.4).

It is important to emphasize that more performance indicators than those listed in Table 8.1 can be improved if the right derivatization reaction is chosen. Detection selectivity and separation selectivity are not listed in Tables 5.1–5.4, however these are important aspects of all analytical methods. These parameters are related to the ability of a method to get rid of substances that can interfere with the analysis. Therefore, a method with a good selectivity will produce chromatograms and electropherograms that are clean around the bands or peaks of the analytes of interest, resulting in high repeatability and reproducibility, regardless of the concentrations of the interferants (this is discussed in Section 8.4.3).

8.4.1 Fluorogenic Reactions

Table 8.2 gives a list of non-fluorescent reagents that react specifically with primary amines producing fluorescent derivatives. Some properties of OPA (o-Phthaldialdehyde), NDA (Naphthalene-2,3-dicarboxaldehyde), Fluorescamine, FQ (3-(2-Furoyl)quinoline-2-carboxaldehyde), and CBQCA (3-(4-Carboxybenzoyl)quinoline-2-carboxaldehyde) are shown. The reagent fluorescamine (F) has a low fluorescence at the wavelength indicated (385 nm); however any excess reagent will undergo to hydrolysis in the aqueous reaction media within 5 min, resulting in non-fluorescent sub-products. Therefore, the excess of fluorescamine also does not produce extra peaks in the electropherograms.

Most of the primary amines are protonated at both neutral and acidic pHs. Therefore, reactions must be performed in environments that are slightly basic. The parameter Δq shows the charge change of the derivative compared of the original amine at a neutral pH.

Table 8.2 Examples of non-fluorescent reagents that undergo fluorogenic reactions with primary amines to produce fluorescent derivatives. The wavelengths of absorption maximum, emission maximum, and change of charge upon derivatization are shown. These maxima may shift by a few nanometers depending on the target amine and the buffer used in the separation medium.

	Absorption maximum	Emission maximum	Δq
Acronym	nm	nm	e
OPA	330	450	−1
Fluorescamine	385	470	−2
NDA	410-445	460-480	−1
CBQCA	476	550	−2
FQ	480	590	−1

The advantages of fluorogenic derivatization reactions compared to labeling reactions is that they produce clean electropherograms without extra peaks due to the excess of reagent. However the number of fluorogenic reagents is very low compared to labeling reagents and most of them are limited to primary amines as the target functional groups. This is unfortunate, since these fluorogenic reaction greatly improve the detection selectivity.

Attention should also by paid to the changes of the maximum absorption and emission wavelengths when using these reactions, as these maximums are sensitive to both the structure of the target molecule and the separation buffer used, as illustrated by Figure 8.2. Moreover, some of the derivatives, such as OPA, NDA and FQ, are not very soluble in aqueous solutions. Consequently, organic modifiers are recommended to keep the derivatives soluble, both in the reaction tube and during the separation. The black curve of Figure 8.3, for instance, is a sign of the precipitation of the ammonium/ammonia derivative with NDA.

The derivatization reactions of figure 8.2 were performed by adding 5 µL of the target molecule solution (10 mM in 0.1 M borate buffer at pH 9) into a 1.5 mL Eppendorf tube. Next 5 µL of 10 mM fluorescamine in acetonitrile was added, followed by 40 µL of 0.1 M borate buffer at pH 9. The content of the tube was immediately mixed and left to react at room temperature for 10 min before the spectra were recorded.

The derivatization reactions of figure 8.3 were performed by adding 2 µL of the target molecule solution (10 mM in 0.1 M borate buffer at pH 9) into a 1.5 mL Eppendorf tube. Next 10 µL of 10 mM KCN (in water with pH adjusted to 9) was added, mixed, and left at rest for 10 min. 2 µL of 10 mM NDA in acetonitrile was added, followed by 4.7 µL of acetontrile, and 1.5 µL of 0.1 M borate buffer at

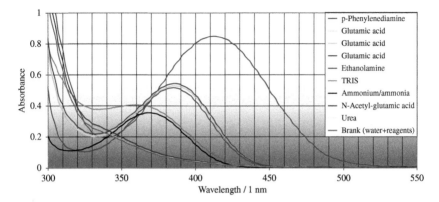

Figure 8.2 Absorption spectra of the products of the fluoregenic reaction of some molecules (mostly amines) with fluorescamine. These molecules are important in the ESTs because some of them are used as buffers while others frequently occur as analytes. The spectra were measured using the "NanoDrop One" UV–vis spectrophorometer and derivatization reactions were conducted in 0.1 M borate buffer pH 9 and using 10 mM fluorescamine in acetonitrile (unpublished results from the author).

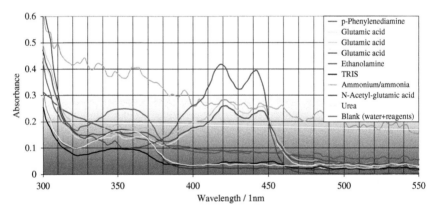

Figure 8.3 Absorption spectra of the products of the fluorogenic reaction of some molecules (mostly amines) with NDA (naphthalene-2,3-dicarboxaldehyde). The irregular black line is a signature of the precipitation of the hydrophobic NDA derivative of ammonium/ammonia. The spectra were measured using the "NanoDrop One" UV–vis spectrophorometer and derivatization reactions were conducted in 0.1 M borate buffer pH 9 and using 2 mM NDA in acetonitrile (unpublished results from the author).

pH 9. The content of the tube was immediately mixed and left to react at room temperature for 1 h 30 min before the spectra were recorded. This derivatization of ammonium/ammonia with NDA produced a pink precipitate that was visually noticeable after 10 min of reaction.

8.4.2 Labeling Reactions

The labeling reactions of Figure 8.4 were performed as follows: 2 µL of the target molecule solution (10 mM in 0.1 M borate buffer at pH 9) was added into a 1.5 mL Eppendorf tube. 10 µL of 2 mM 7-hydroxycoumarin-3-carboxylic acid N-succinimidyl ester in DMSO was added, followed by 8 µL 0.1 M sodium borate buffer pH 9, mixed, and left to react at room temperature for 1 h 30 min before the spectra were taken.

Figure 8.4 shows that there are fewer changes in the absorption maxima of the labeled derivatives than there are in the absorption maxima of the fluorogenic derivatives (Figures 8.2 and 8.3). This occurs because the target analyte participates with much more bonds in the fluorogenic reaction than in the labeling reaction. However, the excess of reagents in the fluorogenic reactions is non-fluorescent, while the products of the reactions are fluorescent. It is very advantageous if the excess of reagents is non-fluorescent, as they will not appear in the electropherograms and chromatograms.

8.4.3 Improving Selectivity Through Derivatization

Separation selectivity and detection selectivity are two different properties that are always desired in a method. Derivatization reactions can improve the former, the

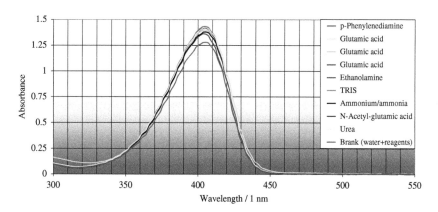

Figure 8.4 Absorption spectra of compounds (mostly amines) labeled with 7-hydroxycoumarin-3-carboxylic acid N-succinimidyl ester (HC-SE). Note that the label is free (unbound) following its attempted reaction with urea, blank, and N-acetyl-glutamic acid, as these reagents do not react with the reactive label moiety (succinimidyl ester). The absorption maxima of the products (labeled and unlabeled) do not change very much in this case. The spectra were measured using the "NanoDrop One" UV-vis spectrophorometer and derivatization reactions were conducted in 0.1 M borate buffer pH 9 and using 2 mM HC-SE in acetonitrile (unpublished results from the author).

latter, and sometimes even both. The goal of this subsection is to describe some of these possibilities.

Many biological matrices are very complex and contain a large number of potential interferants. One way to overcome this, when trying to analyse a few analytes in a complex mixture, is to use brute force, i.e. multiple separation steps are used to push the separation efficiency to extremely high values. However, this is costly and time demanding. An alternative, interesting way to overcome this is to use derivatization.

Separation selectivity can be improved when a derivatization reagent that gives a reaction product which has a mobility characteristic that is significantly different from all matrix constituents (potential interferants) is used. One such elegant procedure is the derivatization of reducing sugars with 8-aminopyrene-1,3,6-trisulfonic acid. The trisulfonate pyrene label adds a strong anodic mobility to the derivatives upon conjugation, improving the separation selectivity and many of the performance indicators shown in Table 5.4.

Detection selectivity is related to the detection and not to the separation mechanism itself. The separation is still important, but the derivatizing reagent is chosen in such a way that it reacts with a functional group, which is common among the components of interest (analytes) and rare among all other possible constituents of the sample. In this case a much smaller number of peaks will appear on the electropherogram because the great majority of the sample constituents pass the detection point without being recorded. To illustrate this let us suppose that free L-proline and L-hydroxyproline must be analyzed in a complex sample such as animal muscle tissue. This kind of sample contains a great variety of molecules including metabolites, amino acids, peptides, proteins, amino acid derivatives, and so on. Figure 8.5 shows such an example. A sample of canned beef was mixed with distilled water and homogenized using a strong mixer to disrupt the tissues. It was then centrifuged and the aqueous phase (intermediate phase) filtered by using an Amicon tube with 10000 Da cut-off membrane using centrifuge at 14000g until 500 μl of filtrated material was produced. After this the sample is subject to react first with orto-phthaldihaldehyde (absorption maximum ~ 337 nm), to block all primary amines. This is followed with a derivatization reaction with 7-aminoethylcoumarin isothiocyanate (absorption maximum ~ 400 nm). The fluorescent end product are primary amine derivatives with absortion maximum at 337 nm and secondary amine derivatives with absorption maximum at 400 nm. Therefore, only the secondary amine derivatives (derivatives of L-proline, D-proline, L-hydroxyproline, and D-hydroxyproline) will be detected and all others will cause no extra detected bands and peaks if excitation at 400 nm is used. Figure 8.5 illustrates such an example.

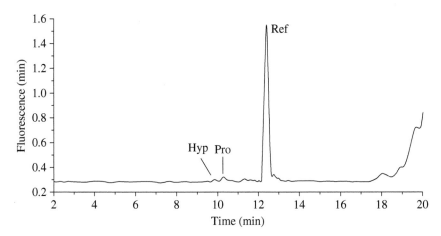

Figure 8.5 Analysis of free proline and hydroxyproline in canned beef. The sample was mixed with distilled water using a mixer, centrifuged, and the aqueous phase (intermediate phase) was filtered using an Amicon tube with 10000 Da cut-off membrane – achieved by centrifuge at 14000g until 500 μl of filtrated material was produced. After a double-step derivatization the sample was hydrodynamically injected and analyzed using MEKC (25 mM borate buffer with 50 mM SDS and pH=9), 15 kV applied voltage, and a capillary of 60 cm total length and 55 cm effective length (author's unpublished results).

References

1 Kubáň, P. and Hauser, P.C. (2018). *TrAC Trends in Analytical Chemistry* 102: 311–321.
2 Galievsky, V.A., Stasheuski, A.S., and Krylov, S.N. (2016). *Analytica Chimica Acta* 935: 58–81.
3 Ban, E. and Song, E.J. (2013). *Journal of Chromatography B: Analytical Technologies in the Biomedical and Life Sciences* 929: 180–186.
4 Kárník, K. (2015). *Electrophoresis* 36: 159–178.
5 Kárník, K. (2013). *Electrophoresis* 34: 70–85.
6 Vessman, J., Stefan, R.I., van Staden, J.F. et al. (2001). *Pure and Applied Chemistry* 73: 1381–1386.
7 IUPAC (2012). Compendium of Chemical Terminology, Gold Book, Version 2.3.2. Research Triangle Park, NC: International Union of Pure and Applied Chemistry.
8 Qu, Y. and Dovichi, N.J. (2018). *TrAC Trends in Analytical Chemistry* 108: 23–37.
9 Beale, S.C. and Sudmeier, S.J. (1995). *Analytical Chemistry* 67: 3367–3371.

10 Yao, B., Yang, H., Liang, Q. et al. (2006). *Analytical Chemistry* 78: 5845–5850.

11 Lo, R.C. and Ugaz, V.M. (2008). *Lab on a Chip* 8: 2135–2145.

12 Zeng, H., Glawdel, T., and Ren, C.L. (2015). *Electrophoresis* 36: 2542–2545.

13 Underberg, W.J.M. and Waterval, J.C.M. (2002). *Electrophoresis* 23: 3922–3933.

14 Waterval, J.C.M., Lingeman, H., Bult, A., and Underberg, W.J.M. (2000). *Electrophoresis* 21: 4029–4045.

15 Bardelmeijer, H.A., Lingeman, H., De Ruiter, C., and Underberg, W.J.M. (1998). *Journal of Chromatography A* 807: 3–26.

9

Applications of Toroidal Electrophoresis

9.1 Introduction

Open (common) ESTs have an enormous quantity of applications in the microchip, capillary, and slab platforms. Therefore, it would be difficult to describe them all in a single book. Readers are advised to consult review articles to find the best platform and separation mode for each application where the open layout will be used.

Only a few applications have currently been proposed and published for the Toroidal layouts of the ESTs, as this is still not a widely known layout. However, all technological bottlenecks have already been overcome and its only remaining restrictions are: (a) Toroidal capillary electrophoresis, which is the platform that exhibits the highest resolution and peak capacity, requires the fabrication of a fused silica torus from a single piece of capillary using the fusion splicing technique.[1] Most research groups are yet not familiar with this quartz and glass processing technique. (b) The fabrication of the ideal, conic shaped microholes requires some skill or investment in expensive equipment (focused ion beam (FIB) etching or laser drilling). (c) Spatial scanning detectors,[2] which are ideal for maximizing the band capacity in toroidal layouts and to reduce analysis time in open layouts (see Figures 2.8, 2.9, and 2.10), are not commonly used in toroidal or open ESTs. In other words, the spatial scanners allow many more bands to be detected in the toroidal layout due to the limited length between microholes and the high voltage rotation. Fixed point detectors are simpler and are expected to give more reproducible results. They are still good at detecting a set of synchronized components that co-migrate with almost the same mobility and, therefore, are hard to separate. This is a good case for analysis by toroidal electrophoresis. (d) For slabs, the fabrication of chambers with a toroidal geometry that optimizes temperature control is difficult. However, overcoming this problem will be very beneficial as toroidal slab electrophoresis has the potential to deliver the highest separation efficiency out of all of the preparative techniques. (e) Programmable high voltage

Open and Toroidal Electrophoresis: Ultra-High Separation Efficiencies in Capillaries, Microchips, and Slabs, First Edition. Tarso B. Ledur Kist.
© 2020 John Wiley & Sons Ltd. Published 2020 by John Wiley & Sons Ltd.

distributors are not commercially available for any platform. They are a crucial accessory for all ESTs with a toroidal layout.

There are only a few applications of toroidal microchip electrophoresis and toroidal capillary electrophoresis in the literature, as this technology is still under development. To the best of the author's knowledge there are no published applications of toroidal slab electrophoresis, although some unpublished applications do exist.

Toroidal microchip electrophoresis was used in the pioneering work of Burggraf et al., [3] to separate fluorescein from its impurities. They used a 2 cm, square, closed loop path (each side was 0.5 cm long). A total number of six detections and five turns (cycles) was necessary to separate five impurities from fluorescein. The authors observed that the number of theoretical plates increased linearly with the number of turns (or cycles), in accordance with equation 4.8. They also demonstrated that 1.2 million theoretical plates can be obtained from 20

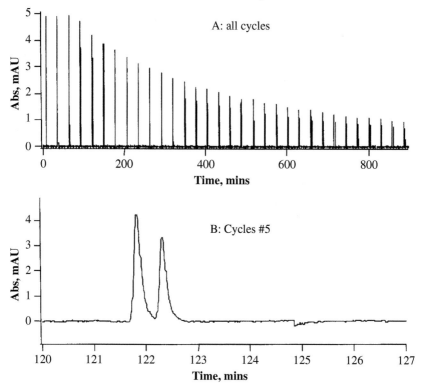

Figure 9.1 Separation of L-phenylalanine from L-phenylalanine-ring-D5 using toroidal capillary micellar electrokinetic chromatography (nomenclature is explained in Appendix A). (A) All 32 cycles are shown. (B) The almost baseline resolution of the two compounds that was achieved in less than 123 min. (*Source*: Reproduced with permission of John Wiley & Sons).

turns within 22 min and 15 s. At the time of publication, the authors called this "synchronous cyclic capillary electrophoresis".

Manz et al. [4] separated FITC-labelled amino acids using free solution electrophoresis (Tris-borate-EDTA) in closed loop microchannels arranged in triangular, square, and pentagonal geometries. They also separated a mixture of dsDNA ladders using the sieving electrophoresis separation mode with fluorescence detection and the aid of the Syber Green I intercalating fluorophore.

Zhao, Hooker, and Jorgenson [5] and Zhao and Jorgenson [6] demonstrated that toroidal capillary electrophoresis can be used to separate the mixture of isotopomers L-phenylalanine and L-phenylalanine-ring-D5 from each other. Baseline resolution was achieved after only five cycles (less than 123 min) using a 2 m long torus with a 50 μm ID, 365 μm OD, and the MEKC separation mode (see Figure 9.1). Technically, the separation platform they used was not really a torus because it was made from four capillaries (~ 49.5 cm each) that were lined up end to end and joined to produce a 2 m closed loop path.

The compound L-phenylalanine-ring-D5 of Figure 9.1A was used to calculate the number of theoretical plates along the run shown in Figure 9.1A. A value of almost 100 million theoretical plates was achieved in this run (shown in Figure 9.2). This has remained a world record for almost two decades.

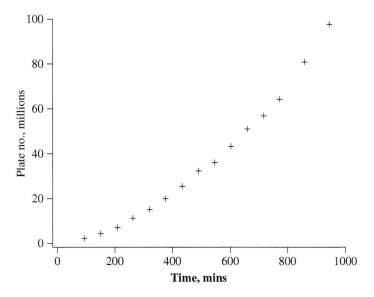

Figure 9.2 Number of theoretical plates calculated from L-phenylalanine-ring-D5 of Figure 9.1. A number of almost 100 million theoretical plates was obtained in less than 900 min. This is the largest number of theoretical plates ever recorded. (*Source:* Reproduced with permission of John Wiley & Sons).

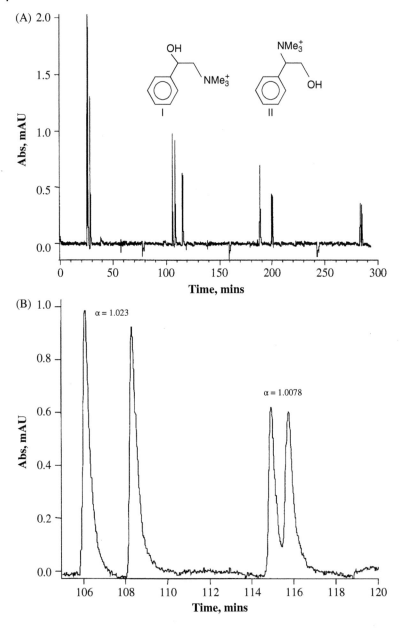

Figure 9.3 Separation of (α-hydroxybenzyl)methyltrimethylammonium and (2-hydroxy-1-phenyl)ethyltrimethylammonium from each other and into their individual stereoisomers using toroidal capillary affinity electrophoresis (nomenclature is explained in Appendix A). (A) The four cycles are shown. Two enantiomers (peaks) of compound II are present at the second detection, but one is lost (electrodriven into the reservoir) by the third detection and both are lost by the fourth detection. (B) Cycle 2 depicts the almost baseline resolution of the four compounds of the mixture that was achieved in less than 123 min. (*Source*: Reproduced with permission of John Wiley & Sons).

The diastereoisomers (α-hydroxybenzyl)methyltrimethylammonium and (2-hydroxy-1-phenyl)ethyltrimethylammonium have been separated from each other and then further separated into their individual stereoisomers.[5,6] Figure 9.3A shows the cycles of this separation, and the inset of this figure shows the structural formula of the two diastereoisomers. Figure 9.3B is a magnification of cycle 2 of Figure 9.3A. It shows the separation of the two diastereoisomers from each other and the separation of each racemate into their component enantiomers.

References

1 Kist, T.B.L. (2017). *Journal of Separation Science* 40: 4619–4627.

2 Beale, S.C. and Sudmeier, S.J. (1995). *Analytical Chemistry* 67: 3367–3371.

3 Burggraf, N., Manz, A., Verpoorte, E. et al. (1994). *Sensors and Actuators B: Chemical* 20: 103–110.

4 Manz, A., Bousse, L., Chow, A. et al. (2001). *Fresenius Journal of Analytical Chemistry* 371: 195–201.

5 Zhao, J., Hooker, T., and Jorgenson, J.W. (1999). *Journal of Microcolumn Separations* 11: 431–437.

6 Zhao, J. and Jorgenson, J.W. (1999). *Journal of Microcolumn Separations* 11: 439–449.

Appendix A

Nomenclature

An overview of the nomenclature has already been given in the Introduction, but a more thorough and concise scheme is proposed here. These names and respective acronyms are an extension of the IUPAC recommendations [1] and the pioneering work of Knox [2]. They follow an intuitive and simple rule: the first word specifies the layout (open or toroidal, although it can be omitted for the "open" layout), the second word indicates the platform on which the electrodriven separation is performed (capillary, microchip or slab) and the remaining term(s) expresses the separation mode used. Overall: name = **layout + platform + Modality**.

	Capillary platform
CAE	(Open) capillary affinity electrophoresis
CEC	(Open) capillary electrochromatography
CELFSE	(Open) capillary end-labeled free-solution electrophoresis
CFSE	(Open) capillary free-solution electrophoresis
CIEF	(Open) capillary isoelectric focusing
CITP	(Open) capillary isotachophoresis
CMEEKC	(Open) capillary microemulsion electrokinetic chromatography
CMEKC	(Open) capillary micellar electrokinetic chromatography
CSE	(Open) capillary sieving electrophoresis
TCAE	Toroidal capillary affinity electrophoresis
TCEC	Toroidal capillary electrochromatography
TCELFSE	Toroidal capillary end-labeled free-solution electrophoresis
TCFSE	Toroidal capillary free-solution electrophoresis
TCIEF	Toroidal capillary isoelectric focusing[a]
TCITP	Toroidal capillary isotachophoresis[b]

(Continued)

Open and Toroidal Electrophoresis: Ultra-High Separation Efficiencies in Capillaries, Microchips, and Slabs, First Edition. Tarso B. Ledur Kist.
© 2020 John Wiley & Sons Ltd. Published 2020 by John Wiley & Sons Ltd.

TCMEEKC	Toroidal capillary microemulsion electrokinetic chromatography
TCMEKC	Toroidal capillary micellar electrokinetic chromatography
TCSE	Toroidal capillary sieving electrophoresis

Microchip platform

MAE	(Open) microchip affinity electrophoresis
MEC	(Open) microchip electrochromatography
MELFSE	(Open) microchip end-labeled free-solution electrophoresis
MFSE	(Open) microchip free-solution electrophoresis
MIEF	(Open) microchip isoelectric focusing
MITP	(Open) microchip isotachophoresis
MMEEKC	(Open) microchip microemulsion electrokinetic chromatography
MMEKC	(Open) microchip micellar electrokinetic chromatography
MSE	(Open) microchip sieving electrophoresis
TMAE	Toroidal microchip affinity electrophoresis
TMEC	Toroidal microchip electrochromatography
TMELFSE	Toroidal microchip End-Labeled free-solution electrophoresis
TMFSE	Toroidal microchip free-solution electrophoresis
TMIEF	Toroidal microchip isoelectric focusing[a]
TMITP	Toroidal microchip isotachophoresis[b]
TMMEEKC	Toroidal microchip microemulsion electrokinetic chromatography
TMMEKC	Toroidal microchip micellar electrokinetic chromatography
TMSE	Toroidal Microchip sieving electrophoresis

Slab platform

SAE	(Open) slab affinity electrophoresis
SEC	(Open) slab electrochromatography[c]
SELFSE	(Open) slab end-labeled free-solution electrophoresis
SFSE	(Open) slab free-solution electrophoresis
SIEF	(Open) slab isoelectric focusing
SITP	(Open) slab isotachophoresis
SMEEKC	(Open) slab microemulsion electrokinetic chromatography
SMEKC	(Open) slab micellar electrokinetic chromatography
SSE	(Open) slab sieving electrophoresis
TSAE	Toroidal slab affinity electrophoresis
TSEC	Toroidal slab electrochromatography[c]
TSELFSE	Toroidal slab end-labeled free-solution electrophoresis

TSFSE	Toroidal slab free-solution electrophoresis
TSIEF	Toroidal slab isoelectric focusing[a]
TSITP	Toroidal slab isotachophoresis[b]
TSMEEKC	Toroidal slab microemulsion electrokinetic chromatography
TSMEKC	Toroidal slab micellar electrokinetic chromatography
TSSE	Toroidal slab sieving electrophoresis

a) These techniques are only viable during the last round trip of the analytes when multidimensional toroidal electrophoresis is used.
b) These techniques are unlikely to be feasible due to the mechanism of the separation mode.
c) These techniques are unlikely to be advantageous due to the poor heat dissipation observed in this platform.

Sometimes the separation mode does not need to be specified. In these cases the names only take into account the layout and the platform:

Acronym	Layout and platform
CE	(Open) capillary electrophoresis
ME	(Open) microchip electrophoresis
SE	(Open) Slab electrophoresis
TCE	Toroidal capillary electrophoresis
TME	Toroidal microchip electrophoresis
TSE	Toroidal Slab electrophoresis

These can be regarded as the six main types of instruments, since the separation mode only depends on the separation matrix used in the platform (capillary, microchip or slab). On the other hand, some separation modes (e.g., entangled polymer solutions) require special changes to be made to the hardware (e.g., high pressure syringes) and dedicated software as well. As a result it is common to see instruments dedicated to a specific layout, platform, and separation mode.

Free-flow electrophoresis (FFE) is a special layout for preparative separations that received this name in the middle of the last century [3,4]. It works in a two-dimensional platform and many separation modes can be used. In this setup, the sample is continuously fed into a laminar buffer stream flow as a narrow streak into the stream along a two-dimensional structure. As illustrated by Figure A.1, an electric field is applied orthogonally to this flow direction and parallel to the two-dimensional chamber [5]. The analytes of interest are collected at specific points at the stream outlet. This FFE is applied for the preparative separation and collection of cell organelles [6], proteins in proteomes [7], peptides

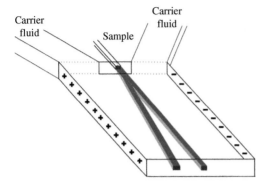

Figure A.1 Illustration of the setup of a typical free flow electrophoresis device, showing the continuous feeding of the sample and the continuous buffer flow. (*Source*: Reproduced with permission of American Chemical Society.)

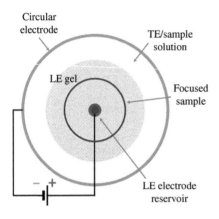

Figure A.2 Illustration of the epitachophoresis setup recently described. It belongs to open slab isotachophoresis with a special geometry (circular). The samples are applied externally with a ring format, migrate inwards, and can be collected at the center. (*Source*: Reproduced with permission of American Chemical Society.)

[8], graphene oxides [9], plant golgi apparatus and endoplasmic reticulum [10], bacteria [11], DNA aptamers [12], and nanoparticles [13], to cite a few. The sample is hydrodynamically fed at constant velocity in a direction that is orthogonal to the direction of electrophoretic separation. Therefore, it falls within the category of open slab electrophoresis with a special geometry and operation.

   Another interesting example to be examined is epitachophoresis [14], an analytical and preparative technique recently described that also allows large amounts of separated fractions to be collected. According to the above nomenclature, this is a special geometry of open slab isotachophoresis. The sample is applied in a ring well and migrates inwards, as shown in Figure A.2. This technique belongs to the open layout because the analytes cannot recirculate. It belongs to the slab platform because it uses an agarose disk. Finally, it belongs to the isotachophoresis separation mode because it uses a leading electrolyte and a terminating electrolyte.

References

1 Riekkola, M.-L., Jönsson, J.Å., and Smith, A.R.M. (2004). *Pure and Applied Chemistry* 76: 443–451.

2 Knox, J.H. (1994). *Journal of Chromatography A* 68: 3–13.

3 Kolin, A. (1965). *Journal of Chromatography A* 17: 532–537.

4 Sulkowski, E. and Laskowski, M. (1967). *Analytical Biochemistry* 20: 94–101.

5 Saar, K.L., Müller, T., Charmet, J. et al. (2018). *Analytical Chemistry* 90: 8998–9005.

6 Islinger, M., Wildgruber, R., and Völkl, A. (2018). *Electrophoresis* 39: 2288–2299.

7 Wildgruber, R., Weber, G., Wise, P. et al. (2014). *Proteomics* 14: 629–636.

8 Xia, Z.-J., Liu, Z., Kong, F.-Z. et al. (2017). *Electrophoresis* 38: 3147–3154.

9 Liu, Y., Zhang, D., Pang, S. et al. (2015). *Journal of Separation Science* 38: 157–163.

10 Parsons, H.T., Fernández-Niño, S.M.G., and Heazlewood, J.L. (2014). *Methods in Molecular Biology* 1072: 527–539.

11 Podszun, S., Vulto, P., Heinz, H. et al. (2012). *Lab on a Chip* 12: 451–457.

12 Jing, M. and Bowser, M.T. (2011). *Lab on a Chip* 11: 3703–3709.

13 Ho, S., Critchley, K., Lilly, G.D. et al. (2009). *Journal of Materials Chemistry* 19: 1390–1394.

14 Foret, F., Datinská, V., Voráčová, I. et al. (2019). *Analytical Chemistry* 91: 7047–7053.

Appendix B

Species Concentration in Buffer Solutions

To prepare a buffer solution the weak acid H_nA is dissolved in water and the pH is adjusted using a strong base, which we will consider to be completely dissociated. Conversely, a buffer can be prepared by using a weak base B and adding a strong acid to adjust the pH (this strong acid will also be considered completely dissociated). It is important to have the concentration profile of the species as a function of pH, as many species interact with analytes while others exhibit different spectroscopic (absorption and fluorescence) properties.

In this book we will stay within the Brønsted–Lowry theory of acid–base reactions. According to this theory there are up to $n+1$ species that result from the ionization of the weak acid H_nA or the conjugated acid BH_n^{n+} of the weak base B. The variable n may assume the value of $n=1$ for monoprotic acids, $n=2$ for diprotic acids, $n=3$ for triprotic acids, and $n=4$ for tetraprotic acids. At certain pHs these $n+1$ species may coexist at similar concentrations. Therefore, these relationships may be very useful in the field of electrodriven separations.

The following equations give the concentrations of each of these species as a function of the following variables: pH of the buffer solution, the initial concentration amount c^{tot} (molar) of the acid H_nA (or base B) used to prepare the buffer solution and the pK_{ai} of each acid (or conjugated acids of base B). They are derived from a set of linear equations, which are laborious to solve by hand but are easily obtained using symbolic computing environment software. These same equations are given in the Log-Log format by Kahlert and Scholz [1]. A detailed discussion of acid–base equilibria is given by Connors [2]. Assuming that the activity coefficients (γ_i) are close to unity ($\gamma_i \simeq 1$, see equation 1.10), then the following equations are obtained.

Open and Toroidal Electrophoresis: Ultra-High Separation Efficiencies in Capillaries, Microchips, and Slabs, First Edition. Tarso B. Ledur Kist.
© 2020 John Wiley & Sons Ltd. Published 2020 by John Wiley & Sons Ltd.

B.1 Acids (H_nA)

B.1.1 Monoprotic Acids ($n = 1$)

$$c_{\mathrm{HA}} = c_{\mathrm{HA}}^{\mathrm{tot}}[1 + 10^{\mathrm{pH}-\mathrm{p}K_{a1}}]^{-1} \tag{B.1}$$

$$c_{\mathrm{A^-}} = c_{\mathrm{HA}}^{\mathrm{tot}}[1 + 10^{-\mathrm{pH}+\mathrm{p}K_{a1}}]^{-1}. \tag{B.2}$$

B.1.2 Diprotic Acids ($n = 2$)

$$c_{\mathrm{H_2A}} = c_{\mathrm{H_2A}}^{\mathrm{tot}}[1 + 10^{\mathrm{pH}-\mathrm{p}K_{a1}} + 10^{2\mathrm{pH}-\mathrm{p}K_{a1}-\mathrm{p}K_{a2}}]^{-1} \tag{B.3}$$

$$c_{\mathrm{HA^-}} = c_{\mathrm{H_2A}}^{\mathrm{tot}}[1 + 10^{-\mathrm{pH}+\mathrm{p}K_{a1}} + 10^{\mathrm{pH}-\mathrm{p}K_{a2}}]^{-1} \tag{B.4}$$

$$c_{\mathrm{A^{2-}}} = c_{\mathrm{H_2A}}^{\mathrm{tot}}[1 + 10^{-2\mathrm{pH}+\mathrm{p}K_{a1}+\mathrm{p}K_{a2}} + 10^{-\mathrm{pH}+\mathrm{p}K_{a2}}]^{-1}. \tag{B.5}$$

B.1.3 Triprotic Acids ($n = 3$)

$$
\begin{aligned}
c_{\mathrm{H_3A}} = c_{\mathrm{H_3A}}^{\mathrm{tot}}[1 &+ 10^{\mathrm{pH}-\mathrm{p}K_{a1}} + 10^{2\mathrm{pH}-\mathrm{p}K_{a1}-\mathrm{p}K_{a2}} \\
&+ 10^{3\mathrm{pH}-\mathrm{p}K_{a1}-\mathrm{p}K_{a2}-\mathrm{p}K_{a3}}]^{-1}
\end{aligned}
\tag{B.6}
$$

$$
\begin{aligned}
c_{\mathrm{H_2A^-}} = c_{\mathrm{H_3A}}^{\mathrm{tot}}[1 &+ 10^{\mathrm{pH}-\mathrm{p}K_{a1}} + 10^{\mathrm{pH}-\mathrm{p}K_{a2}} \\
&+ 10^{2\mathrm{pH}-\mathrm{p}K_{a2}-\mathrm{p}K_{a3}}]^{-1}
\end{aligned}
\tag{B.7}
$$

$$
\begin{aligned}
c_{\mathrm{HA^{2-}}} = c_{\mathrm{H_3A}}^{\mathrm{tot}}[1 &+ 10^{-2\mathrm{pH}+\mathrm{p}K_{a1}+\mathrm{p}K_{a2}} + 10^{-\mathrm{pH}+\mathrm{p}K_{a2}} \\
&+ 10^{\mathrm{pH}-\mathrm{p}K_{a3}}]^{-1}
\end{aligned}
\tag{B.8}
$$

$$
\begin{aligned}
c_{\mathrm{A^{3-}}} = c_{\mathrm{H_3A}}^{\mathrm{tot}}[1 &+ 10^{-\mathrm{pH}+\mathrm{p}K_{a3}} + 10^{-2\mathrm{pH}+\mathrm{p}K_{a2}+\mathrm{p}K_{a3}} \\
&+ 10^{-3\mathrm{pH}+\mathrm{p}K_{a1}+\mathrm{p}K_{a2}+\mathrm{p}K_{a3}}]^{-1}.
\end{aligned}
\tag{B.9}
$$

B.1.4 Tetraprotic Acids ($n = 4$)

$$
\begin{aligned}
c_{\mathrm{H_4A}} = c_{\mathrm{H_4A}}^{\mathrm{tot}}[1 &+ 10^{\mathrm{pH}-\mathrm{p}K_{a1}} \\
&+ 10^{2\mathrm{pH}-\mathrm{p}K_{a1}-\mathrm{p}K_{a2}} + 10^{3\mathrm{pH}-\mathrm{p}K_{a1}-\mathrm{p}K_{a2}-\mathrm{p}K_{a3}} \\
&+ 10^{4\mathrm{pH}-\mathrm{p}K_{a1}-\mathrm{p}K_{a2}-\mathrm{p}K_{a3}-\mathrm{p}K_{a4}}]^{-1}
\end{aligned}
\tag{B.10}
$$

$$
\begin{aligned}
c_{\mathrm{H_3A^-}} = c_{\mathrm{H_4A}}^{\mathrm{tot}}[1 &+ 10^{-\mathrm{pH}+\mathrm{p}K_{a1}} \\
&+ 10^{\mathrm{pH}-\mathrm{p}K_{a2}} + 10^{2\mathrm{pH}-\mathrm{p}K_{a2}-\mathrm{p}K_{a3}} \\
&+ 10^{3\mathrm{pH}-\mathrm{p}K_{a2}-\mathrm{p}K_{a3}-\mathrm{p}K_{a4}}]^{-1}
\end{aligned}
\tag{B.11}
$$

$$
\begin{aligned}
c_{\mathrm{H_2A^{2-}}} = c_{\mathrm{H_4A}}^{\mathrm{tot}}[1 &+ 10^{-2\mathrm{pH}+\mathrm{p}K_{a1}+\mathrm{p}K_{a2}} + 10^{\mathrm{pH}-\mathrm{p}K_{a2}} \\
&+ 10^{\mathrm{pH}-\mathrm{p}K_{a3}} + 10^{2\mathrm{pH}-\mathrm{p}K_{a3}-\mathrm{p}K_{a4}}]^{-1}
\end{aligned}
\tag{B.12}
$$

$$c_{HA^{3-}} = c_{H_4A}^{tot}[1 + 10^{-3pH+pK_{a1}+pK_{a2}+pK_{a3}}$$
$$+ 10^{-2pH+pK_{a2}+pK_{a3}} + 10^{-pH+pK_{a3}}$$
$$+ 10^{pH-pK_{a4}}]^{-1} \tag{B.13}$$

$$c_{A^{4-}} = c_{H_4A}^{tot}[1 + 10^{-4pH+pK_{a1}+pK_{a2}+pK_{a3}+pK_{a4}}$$
$$+ 10^{-3pH+pK_{a2}+pK_{a3}+pK_{a4}} + 10^{-2pH+pK_{a3}+pK_{a4}}$$
$$+ 10^{-pH+pK_{a4}}]^{-1}. \tag{B.14}$$

B.2 Bases (B)

B.2.1 Monoprotonated Bases ($n = 1$)

$$c_{BH^+} = c_B^{tot}[1 + 10^{pH-pK_{a1}}]^{-1} \tag{B.15}$$
$$c_B = c_B^{tot}[1 + 10^{-pH+pK_{a1}}]^{-1}. \tag{B.16}$$

B.2.2 Diprotonated Bases ($n = 2$)

$$c_{BH_2^{2+}} = c_B^{tot}[1 + 10^{pH-pK_{a1}} + 10^{2pH-pK_{a1}-pK_{a2}}]^{-1} \tag{B.17}$$
$$c_{BH^+} = c_B^{tot}[1 + 10^{-pH+pK_{a1}} + 10^{pH-pK_{a2}}]^{-1} \tag{B.18}$$
$$c_B = c_B^{tot}[1 + 10^{-2pH+pK_{a1}+pK_{a2}} + 10^{-pH+pK_{a2}}]^{-1}. \tag{B.19}$$

B.2.3 Triprotonated Bases ($n = 3$)

$$c_{BH_3^{3+}} = c_B^{tot}[1 + 10^{pH-pK_{a1}} + 10^{2pH-pK_{a1}-pK_{a2}}$$
$$+ 10^{3pH-pK_{a1}-pK_{a2}-pK_{a3}}]^{-1} \tag{B.20}$$
$$c_{BH_2^{2+}} = c_B^{tot}[1 + 10^{pH-pK_{a1}} + 10^{pH-pK_{a2}}$$
$$+ 10^{2pH-pK_{a2}-pK_{a3}}]^{-1} \tag{B.21}$$
$$c_{BH^+} = c_B^{tot}[1 + 10^{-2pH+pK_{a1}+pK_{a2}} + 10^{-pH+pK_{a2}}$$
$$+ 10^{pH-pK_{a3}}]^{-1} \tag{B.22}$$
$$c_B = c_B^{tot}[1 + 10^{-pH+pK_{a3}} + 10^{-2pH+pK_{a2}+pK_{a3}}$$
$$+ 10^{-3pH+pK_{a1}+pK_{a2}+pK_{a3}}]^{-1}. \tag{B.23}$$

B.2.4 Tetraprotonated Bases ($n = 4$)

$$c_{BH_4^{4+}} = c_B^{tot}[1 + 10^{pH-pK_{a1}} + 10^{4pH-pK_{a1}-pK_{a2}-pK_{a3}-pK_{a4}}$$
$$+ 10^{2pH-pK_{a1}-pK_{a2}} + 10^{3pH-pK_{a1}-pK_{a2}-pK_{a3}}]^{-1} \tag{B.24}$$

$$c_{BH_3^{3+}} = c_B^{tot}[1 + 10^{-pH+pK_{a1}} + 10^{pH-pK_{a2}} + 10^{2pH-pK_{a2}-pK_{a3}}$$

$$+ 10^{3pH-pK_{a2}-pK_{a3}-pK_{a4}}]^{-1} \tag{B.25}$$

$$c_{BH_2^{2+}} = c_B^{tot}[1 + 10^{-2pH+pK_{a1}+pK_{a2}} + 10^{pH-pK_{a2}} + 10^{pH-pK_{a3}}$$

$$+ 10^{2pH-pK_{a3}-pK_{a4}}]^{-1} \tag{B.26}$$

$$c_{BH^+} = c_B^{tot}[1 + 10^{-3pH+pK_{a1}+pK_{a2}+pK_{a3}} + 10^{-2pH+pK_{a2}+pK_{a3}}$$

$$+ 10^{-pH+pK_{a3}} + 10^{pH-pK_{a4}}]^{-1} \tag{B.27}$$

$$c_B = c_B^{tot}[1 + 10^{-4pH+pK_{a1}+pK_{a2}+pK_{a3}+pK_{a4}}$$

$$+ 10^{-3pH+pK_{a2}+pK_{a3}+pK_{a4}} + 10^{-2pH+pK_{a3}+pK_{a4}}$$

$$+ 10^{-pH+pK_{a4}}]^{-1}. \tag{B.28}$$

References

1 Kahlert, H. and Scholz, F. (2013). *Acid-Base Diagrams*. Berlin-Heidelberg: Springer-Verlag.

2 Connors, K.A. (2003). *Thermodynamics of Pharmaceutical Systems*. Hoboken, NJ: Wiley.

Appendix C

Electrophoresis

C.1 Free-Solution Electrophoretic Mobility

Suppose an ion or electrically charged particle that is placed in a homogeneous liquid, that has a constant dynamic viscosity, and a low Reynolds number (no mass convection and no turbulence). This single ion or charged particle will have the constant charge q, constant mass m, and will be subjected to white noise ($\mathbf{w}$) as well as an applied electric field ($\mathbf{E}$). This system can be modeled by *Langevin's equation* [1,2]:

$$\frac{d\mathbf{v}}{dt} = -\frac{f}{m}\mathbf{v} + \frac{\mathbf{w}}{m} + \frac{q}{m}\mathbf{E}, \tag{C.1}$$

where $\mathbf{v}$ is the velocity. The viscous friction coefficient, due to the dynamic viscosity η, is represented by f, and white noise ($\mathbf{w}$) has the following properties:

$$\langle \mathbf{w}(t) \rangle = 0 \quad \text{and} \quad \langle \mathbf{w}(t)\,\mathbf{w}(t') \rangle = \delta(t - t'), \tag{C.2}$$

where the symbol $\langle\ \rangle$ denotes the average of the possible values that $\mathbf{w}(t)$ may assume at time t, and $\delta(t - t')$ is the *Dirac delta function*.

When a given sequence of random noise is applied to the system during the time interval $[0, t]$, assuming that the electric field is zero for $t < 0$ and $\mathbf{E}$ for $t \geq 0$, then equation C.1 becomes:

$$\mathbf{v}(t) = \frac{q}{f}\mathbf{E} + \left[\mathbf{v}(0) - \frac{q}{f}\mathbf{E}\right] e^{-ft/m} + \frac{1}{m}\int_0^t \mathbf{w}(s)\,e^{-f(t-s)/m}\,ds. \tag{C.3}$$

Now, considering $\mathbf{v}(0) = 0$ and taking the average of all possible sequences of random noises, which is equivalent to average a large number of individual realizations or an ensemble of ions or charged particles, the following is obtained:

$$\langle v(t) \rangle = \frac{q}{f}E\left(1 - e^{-\frac{ft}{m}}\right) \quad \text{or} \quad v_e = \mu_e E\left(1 - e^{-\frac{ft}{m}}\right), \tag{C.4}$$

Open and Toroidal Electrophoresis: Ultra-High Separation Efficiencies in Capillaries, Microchips, and Slabs, First Edition. Tarso B. Ledur Kist.
© 2020 John Wiley & Sons Ltd. Published 2020 by John Wiley & Sons Ltd.

where $\langle v(t) \rangle \equiv v_e$ and $\mu_e = q/f$ was used. Only the x component of $\mathbf{E}$ is non-zero, therefore the y and z components of $\mathbf{v}_e$ are zero. This is the classical equation for the free-solution electrophoretic velocity (v_e) of a band if the ensemble average velocity equals the average velocity of many realizations of a single entity (ion or charged particle). The exponential term represents the electrophoretic transient regime and is characterized by the electrophoretic time constant $\tau_e = m/f$. In the steady state (after the transient regime) the electrophoretic velocity is given by:

$$v_e = \frac{q}{f}E \quad \text{or} \quad v_e = \mu_e E. \tag{C.5}$$

The friction coefficient of ions and particles that are spherical, and therefore have the radius r, is given by $f = 6\pi\eta r$. For these ions or particles:

$$v_e = \frac{q}{6\pi\eta r}E \quad \text{or} \quad v_e = \frac{q}{6\pi\nu\rho r}E, \tag{C.6}$$

where η is the dynamic viscosity of the solvent used, ν is its kinematic viscosity, and ρ is its density. This equation is not valid for diluted polymer solutions because in this case the contribution to the viscosity comes from two sources: the solvent and from, e.g., the long and non-interacting neutral polymers dissolved in the solvent. In this case the electrophoretic mobility of small ions is almost independent of the viscosity of the polymer solution [3].

C.1.1 Classical Trajectories

By taking equation C.3 and assuming that $t \gg m/f$ (following the transient regime), it is possible to calculate the random positions that a single particle will occupy over time when subjected to a sequence of random noise. These positions are calculated using:

$$\mathbf{x}(t) = \mathbf{x}(0) + \frac{q}{f}\mathbf{E}t + \frac{1}{m}\int_0^{t'}\int_0^{t}\mathbf{w}(s)\,e^{-f(t'-s)/m}\,ds\,dt'. \tag{C.7}$$

These trajectories can be simulated by using discrete time intervals Δt and the following calculation routine:

$$x(t + \Delta t) = x(t) + \frac{q}{f}E\Delta t + w_i\Delta t,$$
$$y(t + \Delta t) = y(t) + w_j\Delta t, \quad \text{and}$$
$$z(t + \Delta t) = z(t) + w_k\Delta t, \tag{C.8}$$

where the white noise has the following properties: $\langle w_i \rangle = \langle w_j \rangle = \langle w_k \rangle = 0$ and $\langle w_i^2 \rangle = \langle w_j^2 \rangle = \langle w_k^2 \rangle$ is related to the diffusion coefficient of the band. These equations (C.7 and C.8) represent the classical trajectory of a single particle that is affected by a force field, a drag term, and random noise (the observed phenomena is called Brownian motion).

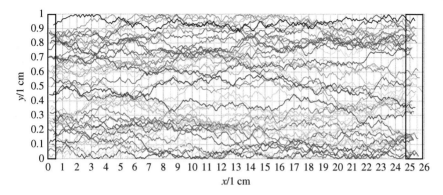

Figure C.1 Simulation of a bi-dimensional free-solution electrophoresis of fifty ions or charged particles that are electrodriven along the x direction according to the stochastic equations C.8. They were injected between $0 < x < 0.5$ cm, using randomly chosen points, and allowed to run for 25 min. The following parameters were used: $v = qE/f = 1$ cm min^{-1}, $\Delta t = 0.1$ min, and white noise was simulated by random numbers with a constant distribution in the $-0.2 < w < 0.2$ range. The left rectangle shows the injected sample plug with $\langle x(0) \rangle = 0.25$ cm and the right rectangle indicates the final positions with $\langle x(25 \text{ min}) \rangle = 25.25$. Note that and $\sigma^2(0) \simeq 0.021$ cm^2, which matches the theoretically predicted value of $(0.5 \text{ cm})^2/12$.

Figure C.1 shows the result of a simulation of a free-solution electrophoresis run of fifty ions or charged particles along a two-dimensional separation path that is 1 cm wide (y direction) and 25 cm long (x direction). The following parameters were used: velocity was $v = 1$ cm min^{-1}, migration time $t_m = 25$ min, noise was represented by random numbers with an even distribution in the range $-0.2 < w < 0.2$, and the initial injected sample plug was 1 cm wide (y direction) and 0.5 cm long (along the separation direction). The position where the sample was injected is indicated by the left rectangle and the final positions of the analytes at time $t = 25$ min is shown by the rectangle at right. The average initial position is $x \simeq 0.25$ cm and the average final position is $x \simeq 25.25$ cm. During the simulation the positions of the ions or charged particles was reflected back when they assumed y values lower than zero or higher than one. These are examples of the classical trajectories of particles subjected to electrophoresis and white noise that would be observed under a microscope.

In some systems the particles cannot be represented by a classical trajectory (positions and velocities) because they exhibit quantum properties, for instance *coherence* and the *tunneling effect*. In these cases their dynamics should be treated within quantum theory, where a *state vector* (or wave-function) is used to represent the ion or particle and *Schrödinger's equation* (usually with a dissipative term) is used to predict its time evolution. When Schrödinger's equation is used the time evolution of a single particle is called *quantum trajectory*, and the average of many trajectories is given by the *master equation* [4]. This is the quantum counterpart of the classical average given by the following equation:

$$\langle x(t) \rangle = x(0) + \frac{qE}{f}t. \tag{C.9}$$

Here, the average positions at time t was calculated over many trajectories. It is assumed that many particles are simultaneously being monitored. Therefore, the movement of the bands in electrophoresis refers to the center of mass of the whole set of ions or charged particles, and not to the maximum of the distribution (band) or the maximum of a peak in an electropherogram. In other words, the migration time is related to the center of mass of the bands and peaks, and not to their maxims.

C.2 Mobility Dependence on Temperature

The dynamic viscosity of liquids decreases with temperature. If the charge of an ion or particle is not affected by temperature, then the dependence of its mobility on temperature can be predicted using the dependence of the solution's mobility dependence on temperature. The dynamic viscosity (η) of pure water, at one atmosphere, is given by:

$$\eta = Ae^{B/(C-T)}. \tag{C.10}$$

where $A = 2.414 \times 10^{-5}$ Pa s, $B = 570.58$ K, $C = 140$ K, and T is the temperature in Kelvin. Figure C.2 plots the behavior of η of pure water in the zero to 100 °C temperature range.

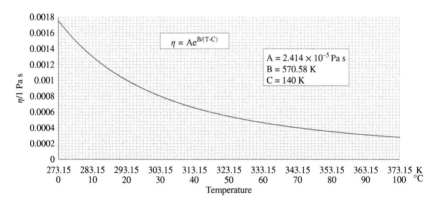

Figure C.2 Plot of equation C.10 showing the dynamic viscosity of pure water in the liquid state from 0 °C (273.15 K) to 100 °C (373.15 K).

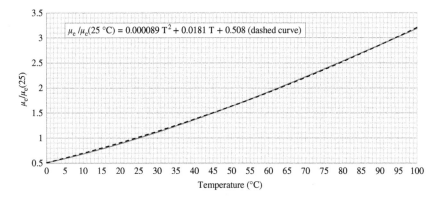

Figure C.3 Plot of mobility ratio against temperature (in Celsius). The second order polynomial gives the best fit for $\mu_e/\mu_e(25\ °C)$. This is exactly equal to $\eta(25\ °C)/\eta$.

It is useful to have the exact electrophoretic mobility of ions and charged particles in pure water as a function of temperature. It is good to take a given temperature, say at 25 °C, and calculate the mobility $\mu_e(T)$ of an ion or charged particle as a function of $\mu_e(25\ °C)$. This result is shown in Figure C.3, where the blue curve represents the exact value of μ_e obtained from equation C.11 and the black dashed line gives the best fit using a second order polynomial. This second order polynomial is given by:

$$\mu_e/\mu_e(25\ °C) = 0.000089\ T^2 + 0.0181\ T + 0.508. \tag{C.11}$$

C.3 Transient Regimes

The transient regimes deserve some attention as they are explored by some of the ESTs, for instance pulsed field electrophoresis. In this separation mode the applied high voltage, and consequently the electric field strength, behave as a periodic function. They are usually square functions, with positive pulses, but sometimes they can even have alternating positive (forward pulses) and negative pulses (backward pulses). One additional degree of freedom can be used in the slab platform due to its two-dimensional nature; the pulses may be applied in different x-y directions by changing the pair of electrodes used at the sides of square, hexagon, and other slab geometries. All this is done to increase the separation efficiency of a selected analyte (usually a macromolecule).

The average electrophoretic velocity of an ion or charged particle that is subjected to an electric field in both the transient and steady state regimes has already been shown in Section C.1. However, there are in fact two different transient regimes:

(1) The time taken by a band made of ions or charged particles to go from rest to the steady state velocity, assuming that the applied electric field goes instantly from zero to its maximum (E). This *electrophoretic transient regime* is given by equation C.4.

(2) The time taken by the output of the high voltage source to go from zero to V volts, i.e. the turn-on time constant or *hardware transient regime*.

C.3.1 Eletrophoretic Transient Regime (τ_e)

There is another simple way, known as the dynamics of the center of mass, to find the electrophoretic mobility and velocity during the electrophoretic transient regime, i.e. when the applied electric field is hypothetically instantly turned on at time $t = 0$ (the applied electric field is zero if $t < 0$, and is equal to E if $t \geq 0$). Considering only the forces with a nonzero average, there are two left acting on the ions and particles: the electric force qE and the opposing friction force fv. Therefore, the average acceleration of an ensemble of ions or charged particles can be given by:

$$\frac{dv}{dt} = \frac{qE - fv}{m} \quad \text{or} \quad \frac{dv}{dt} + \frac{f}{m}v - \frac{qE}{m} = 0. \tag{C.12}$$

The solution to this first order ordinary equation, using the initial condition $v(0) = 0$, results in:

$$v = \frac{qE}{f}\left(1 - e^{-\frac{f}{m}t}\right), \tag{C.13}$$

which is identical to equation C.4. This means that the duration of the transient regime is of the order of the mass-to-friction coefficient ratio.

Figure C.4 shows the transient regime of the center of mass of a band of molecules, assuming that the electric field is instantly turned on at $t = 0$. Note that the velocity of the center of mass reaches ~99% of v_e only when $t/\tau_e = t/(m/f) = 5$, which is the typical behavior of the *increasing form of the exponential decay function*.

C.3.2 Hardware Transient Regime (τ_o)

As the high voltage modules are turned on then the capacitors of their circuitry must be fully charged before the output voltage reaches the set value and, conversely, during the turn off they must discharge completely before the voltage reaches the ground potential. Therefore, instant turn on and turn off is unlikely to be achieved experimentally. The transient voltage, $V(t)$, has almost the same behavior as shown in equation C.13: $V(t) = V(1 - e^{t/\tau_o})$, during turn on period, and $V(t) = Ve^{-t/\tau_o}$ during turn-off, where τ_0 is the time constant of the high voltage source.

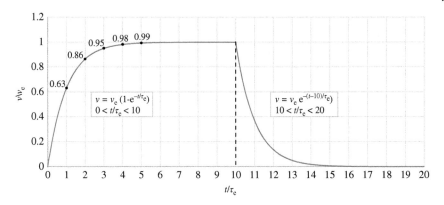

Figure C.4 Velocity of the center of mass of a band during one period of a square pulsed field mode of operation. It is assumed that the applied electric field is zero for $t < 0$ and $t \geq 10$, equal to E for $0 \leq t < 10$, and the duration of one period is $20\tau_e$. Note that the velocity of the center of mass reaches ~99% of v_e only at $t = 5\tau_e$ (the dot at $t/\tau_e = 5$). A time $t = 10$ the electric field is instantly turned off and an exponential decay of the velocity of the center of mass is observed.

If the hardware transient regimes are taken into account, the following is obtained for the electrophoretic velocities of the analytes during the turn on time interval:

$$v_e = -\frac{qE}{f}Ae^{-t/\tau_e} + \frac{qE}{f}(A - B + Be^{-t/\tau_o}), \tag{C.14}$$

where $\tau_e = m/f$ and $A = \tau_e/(\tau_e - \tau_o)$ and $B = \tau_o/(\tau_e - \tau_o)$.

During the turn off interval the velocity of the analyte will change as follows:

$$v_e = \frac{qE}{f}(1 + B - Be^{-t/(B\tau_e)})e^{-t/\tau_e}. \tag{C.15}$$

When a pulsed high voltage with the shape of a square wave, rectangular wave, triangular wave, saw-tooth wave, or any other periodic wave is applied, the hypothetical sharp edges of these periodic functions will always be rounded-off due to the so-called *capacitive impedance* of the high voltage modules. Therefore, in practice, there are no sharp edges in both the applied high voltage or in the applied electric field strength.

All mentioned transient regimes are important for macromolecules, organelles, and cells, which take the time $\tau_e = m/f \geq \tau_o$ to go from rest to the steady state regime. Moreover, the transient regime plays a fundamental role in the ESTs if the time constant τ_e is comparable to the duration of the pulses (square, rectangle, or other). This is used, for instance, to separate dsDNA fragments that are a few hundred kilo-bases long. There are a few reviews on the applications of pulsed-field electrophoresis [5,6], and a few reviews on the fundamentals of pulse field electrophoresis [7].

References

1 Langevin, P. (1908). *Comptes Rendus des Sances de Paris* 146: 530–533.
2 Lemons, D.S. and Gythiel, A. (1997). *American Journal of Physics* 65: 1079–1080.
3 Shimizu, T. and Kenndler, E. (1999). *Electrophoresis* 20: 3364–3372.
4 Kist, T.B.L., Orszag, M., Brun, T.A., and Davidovich, L. (1999). *Journal of Optics B: Quantum and Semiclassical Optics* 1: 251–263.
5 Goering, R.V. (2010). *Infection, Genetics and Evolution* 10: 866–875.
6 Hallin, M., Deplano, A., Denis, O. et al. (2007). *Journal of Clinical Microbiology* 45: 127–133.
7 Cantor, C.R., Smith, C.L., and Mathew, M.K. (1988). *Annual Review of Biophysics and Biophysical Chemistry* 17: 287–304.

Appendix D

Electroosmosis

D.1 Slab and Microchips – Cartesian Coordinates

It is well known [1] that electrostatic potential differences are observed between two distinct phases when they are placed in contact with each other. One such example is dielectric materials that are placed in contact with polar liquids such as aqueous solutions. A net superficial charge density with a particular sign is observed over the surface of the dielectric plane and a diffuse net volumetric charge density with the opposite sign is observed in the liquid phase. Examples of phases that exhibit a surface net charge density when placed in contact with aqueous solutions are: glasses, quartz, polymers, oils, air bubbles, and other inert gases. The resulting surface density depends on the nature of the dielectric plane, the nature of the aqueous solution, and the temperature of the system. A negative net charge density can be observed, for instance, when glasses or fused silica enter into contact with aqueous solutions, as shown in Figure D.1.

The negative charge layer is located exactly at the interface, predominantly as SiO^-, while the positive volumetric net charge density is distributed into the aqueous phase at the expense of thermal energy (expressed by the Boltzmann factor).

The aim of this appendix is to show the shape and depth of the spatial distribution of the positive ions (cations) and the electrostatic potential function from the surface ($y = 0$) all the way into the liquid phase.

The first analytical solution to the electric potential and distribution of charges at these interfaces was obtained by Smoluchowski, [2] Gouy, [3] Chapman [4] and Debye and Hückel. [5]

Hereby we will adopt the model known as the Gouy–Chapman model. Within this model the following assumptions are made:

- The charges inside the liquid phase are considered point-like charges.
- The relative permittivity of the liquid is position-independent, that is, it has the same value across the whole liquid phase.

Open and Toroidal Electrophoresis: Ultra-High Separation Efficiencies in Capillaries, Microchips, and Slabs, First Edition. Tarso B. Ledur Kist.
© 2020 John Wiley & Sons Ltd. Published 2020 by John Wiley & Sons Ltd.

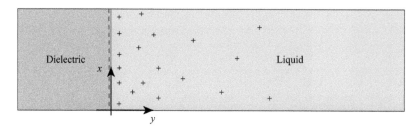

Figure D.1 A schematic representation of the negative surface density of charges and the diffuse volumetric density of positive charges. This is observed, for instance, with glasses and fused silica that is in contact with aqueous solutions.

- The interface is considered perfectly flat, uniformly charged, and infinite.
- The electric work $ze\Psi(y)$ is the only way to bring the charge ze from a point far from the interface, where the potential is zero, to the position y where the potential is $\Psi = \Psi(y)$ ($0 \le y < \infty$).
- The ions are symmetric, that is, the electrolyte can be represented by $(M^{+z}X^{-z})$.
- The distribution of the ions follows the Boltzmann distribution.
- Viscosity is considered constant in the whole liquid (fluid phase).

If the potential is defined as $\Psi(y)$ with $\Psi(0) = \Psi_0$, then applying the Boltzmann law gives:

$$n_+ = n_o \exp\left(\frac{-ze\Psi}{k_B T}\right) \quad \text{and} \quad n_- = n_o \exp\left(\frac{+ze\Psi}{k_B T}\right)$$

where $n_- = n_-(y)$ and $n_+ = n_+(y)$ are the number of ions per unit volume at position y and n_o is the concentration of these ions far from the surface where $\Psi(y) \to 0$. T is the temperature of the system, k_B is the Boltzmann constant, and e is the elementary charge (1.6×10^{-19} C). The net volumetric density of charge will be given by:

$$\rho = ze(n_+ - n_-),$$

$$\rho = z\,e\,n_o \left[\exp\left(\frac{-ze\Psi}{kT}\right) - \exp\left(\frac{+ze\Psi}{kT}\right)\right] \quad \text{and}$$

$$\rho = -2zen_o \sinh\left(\frac{ze\Psi}{kT}\right). \tag{D.1}$$

On the other hand, $\rho(y)$ is related to $\Psi(y)$ through Poisson's equation $\nabla^2\Psi = -\rho/\epsilon$, where ϵ is the electric permittivity of the liquid phase. Using this, equation D.1 and considering that this is a uni-dimensional problem, the following is obtained:

$$\frac{d^2\Psi}{dy^2} = \frac{2zen_o}{\epsilon} \sinh\left(\frac{ze\Psi}{kT}\right). \tag{D.2}$$

This is the Poisson–Boltzmann equation. Using the boundary conditions of $\Psi(0) = \Psi_0$ and $\frac{d\Psi}{dy} = 0$ when $y \to \infty$ (which are widely discussed in the literature [1,6–8]) the following is obtained:

$$\Psi(y) = \frac{2kT}{ze} \ln \left(\frac{1 + \gamma \exp[-\kappa y]}{1 - \gamma \exp[-\kappa y]} \right), \qquad (D.3)$$

where

$$\gamma = \frac{\exp(ze\Psi_0/2kT) - 1}{\exp(ze\Psi_0/2kT) + 1} = \tanh \left(\frac{ze\Psi_0}{4kT} \right)$$

and

$$\kappa = \sqrt{\left(\frac{2z^2 e^2 n_0}{\epsilon kT} \right)} \qquad (D.4)$$

are constants. Parameter κ is known as the inverse of the double-layer thickness, since the extension of the charge distribution and potential is of the order of $1/\kappa$.

The shape of the functions $\Psi(y)$, $n_+(y)$ and $n_-(y)$ are shown in Figure D.2.

If $\frac{ze\Psi_0}{2kT} \ll 1$, then the following approximation can be made:

$$\exp \left(\frac{ze\Psi_0}{2kT} \right) \simeq 1 + \frac{ze\Psi_0}{2kT},$$

and equation D.3, which is the solution of the Poisson–Boltzmann equation D.2), can be simplified to:

$$\Psi \simeq \Psi_0 \exp(-\kappa x), \qquad (D.5)$$

which shows the exponential decay of the electric potential as a function of distance from the surface. Equation D.5 was obtained with the approximation that

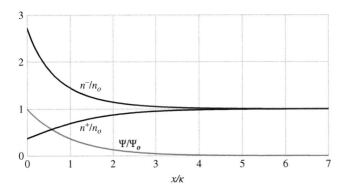

Figure D.2 These curves show the form of the electric potential function and the concentrations of cations and anions for a typical surface with $z = 1$ and $\Psi_0 = -25$ mV.

the potential Ψ_0 is small. For higher potentials ($\Psi_0 \geq 25\,\mathrm{mV}$ for 1-1 electrolytes) this approximation cannot be made.[9]

The surface density of the net charge of σ_0 must be exactly compensated by the net charge of opposite sign in the liquid phase:

$$\sigma_0 = -\int_0^\infty \rho dy \, ,$$

which results in:

$$\sigma_0 = \sqrt{8n_0 \epsilon kT} \sinh \left(\frac{ze\Psi_0}{2kT} \right) \tag{D.6}$$

or:

$$\sigma_0 = \frac{4zn_0 e}{\kappa} \sinh \left(\frac{ze\Psi_0}{2kT} \right) .$$

At low potentials this reduces to:

$$\sigma_0 = \epsilon\kappa\Psi_0 \propto z\sqrt{\frac{n_0}{T}}\Psi_0 . \tag{D.7}$$

These equations show the relationship between σ_0 and Ψ_0.

D.2 Capillaries – Cylindrical Coordinates

The quantitative treatment given in the previous section to obtain $\rho(y)$ and $\Psi(y)$ was exact, because it was considered a flat and infinite surface. These are valid results for chambers used in slab electrophoresis and microchannels in microchips when $1/\kappa$ is much smaller than the width and depth of the microchannels, which is usually the case for all practical applications of ESTs.

In the situation where dielectric materials with cylindrical shapes (capillaries) and a radius of $r = a$ are filled with a liquid, the Poisson–Boltzmann equation D.2 can be rewritten as:

$$\frac{1}{r}\frac{\mathrm{d}}{\mathrm{d}r}\left(r\frac{\mathrm{d}\Psi(r)}{\mathrm{d}r} \right) = -\frac{\rho(r)}{\epsilon} \, , \tag{D.8}$$

where r is the distance from the central symmetry axis. In this case, due to the coordinate system used (cylindrical) and the symmetry of the system, Ψ and ρ depend only on r. Therefore, equation D.1 can be rewritten as:

$$\rho(r) = -2zen_0 \sinh \left(\frac{ze\Psi(r)}{kT} \right) , \tag{D.9}$$

If $\frac{ze\Psi}{kT}$ is small, then:

$$\sinh \left(\frac{ze\Psi}{kT} \right) \approx \frac{ze\Psi}{kT} . \tag{D.10}$$

In this case, equation D.8 simplifies and assumes the following form:

$$\frac{1}{r}\frac{d}{dr}\left(r\frac{d\Psi}{dr}\right) = \kappa^2\Psi,\tag{D.11}$$

This approximation is known as the Debye–Hückel approximation. The solution of D.11 must be finite at $r = 0$, therefore, $\Psi(r) = BI_0(\kappa r)$, where I_0 is the modified Bessel function of first species of zero order [10] and B a constant. If $r = a$ the condition is $\Psi(a) = \Psi_0$, and:

$$\Psi(r) = \Psi_0\frac{I_0(\kappa r)}{I_0(\kappa a)},\tag{D.12}$$

The volumetric density of charges is given by

$$\rho(r) = -\epsilon\kappa^2\Psi(r) = -\epsilon\kappa^2\Psi_0\frac{I_0(\kappa r)}{I_0(\kappa a)}.\tag{D.13}$$

With the expressions of $\Psi(r)$ and $\rho(r)$ in hand, it is now possible to analyze the dynamics of a fluid that fills the lumen of a cylinder when an electrostatic potential gradient is applied along the axial direction (they are used to drive the separations). Moreover, axial pressure gradient (dP/dx) may be also simultaneously applied along this same axial direction. Under these conditions and considering the symmetry of the problem, the other spatial derivatives of V and P are all null in the regime of laminar flow (non-turbulent).

The equation of the movement of a viscous and isotropic liquid (in the linear approximation of the viscous effects with constant values of η and f) is given by:

$$\rho\left[\frac{\partial\vec{v}}{\partial t} + (\vec{v}\cdot\nabla)\vec{v}\right] = -\nabla P + \eta\nabla^2\vec{v} + (f + \eta/3)\nabla(\nabla\cdot\vec{v}).$$

For incompressible fluids $\nabla\cdot\vec{v} = 0$. Adding a force per unit volume $\vec{F}$ due to the net drag force on the liquid that comes from the movement of the ions with density of charge $\rho(r)$ due to the applied electric field $\vec{E}$ results in:

$$\rho\left[\frac{\partial\vec{v}}{\partial t} + (\vec{v}\cdot\nabla)\vec{v}\right] = -\nabla P + \eta\nabla^2\vec{v} - \vec{F}.\tag{D.14}$$

This is the Navier–Stokes equation with an external force term. We are interested in the steady state regime when $\frac{\partial\vec{v}}{\partial t} = 0$. Due to the cylindrical geometry the term $(\vec{v}\cdot\nabla)\vec{v}$ is also null, resulting in:

$$\frac{1}{r}\frac{d}{dr}\left(r\frac{dv_x}{dr}\right) = \frac{1}{\eta}\frac{dP}{dx} - \frac{F_x}{\eta}.\tag{D.15}$$

Using $-\frac{dP}{dx} = P_x$, where P_x is the pressure gradient that is applied uniformly along the the cylinder and F_x is the resulting force of the action of the applied electric field E_x on the charges represented by the distribution $\rho(r)$. Equation D.15 can be rewritten as:

$$\frac{d^2 v_x}{dr^2} + \frac{1}{r}\frac{dv_x}{dr} = -\frac{P_x}{\eta} + \frac{E_x \epsilon \kappa^2 \Psi_0}{\eta}\frac{I_0(\kappa r)}{I_0(\kappa a)} . \tag{D.16}$$

Using the condition that P and v_x must be finite at all points and using:

$$v_x(a) = \left(\frac{dv_x}{dr}\right)_{r=0} = 0 , \tag{D.17}$$

the following is obtained:

$$v_x(r) = \frac{P_x}{4\eta}(a^2 - r^2) - \frac{E_x \epsilon \Psi_0}{\eta}\left[1 - \frac{I_0(\kappa r)}{I_0(\kappa a)}\right] . \tag{D.18}$$

Note that v_x is the sum of two terms: the Poiseuville term and the electrokinetic term. Figure 2.18 shows the velocity profiles inside the capillary for the following situations: (A) only electroosmosis is present, (B) when only a pressure driven flow exist, (C) the EOF and pressure driven co-exist and point to the same direction, and (D) the EOF and pressure driven co-exist and point in opposite directions.

The parameter κa indicates how much larger the inner radius of the capillary is compared to the thickness of the double layer $(1/\kappa)$. This constant is given by equations 2.5 and 2.6, which show that it falls as a decreasing exponential of κy inside the solution (or diffuse layer). For a capillary with a radius of $a = 25$ μm, using $z = 2$, $\epsilon/\epsilon_0 = 80$, $n_0 = 10^{23}$ and $T = 298$ K, a value of $\kappa a = 2096$ is obtained!

As a result, if $P_x = 0$ and $\kappa a \gg 1$ then the profile of velocities is constant practically over the entire lumen of the capillary, except at the points very close to the wall. This "piston-like" flow profile is unique and very interesting for electrodriven micropumps and gives also great advantages to the electrodriven separation techniques because the flow does not promote band mixing like the Poiseuille flow profile observed in chromatography and centrifugation. This constant velocity is equal to the velocity at the center of the capillary and is given by v_{eo}, the electroosmotic velocity, which is expressed as:

$$v_{eo} \equiv v_x(0) \simeq \frac{E_x \epsilon \Psi_0}{\eta} .$$

The electroosmotic mobility (μ_{eo}) is defined as:

$$\mu_{eo} \equiv \frac{v_{eo}}{E_x} \simeq \frac{\epsilon \Psi_0}{\eta} .$$

The goal of this appendix subsection was to show that a laminar flow with a constant velocity is possible under the conditions of $P_x = 0$ and $\kappa a \gg 1$. A laminar flow (without turbulence) is assured when the Reynolds number (R) is smaller than ten $(R < 10)$. Using pure water as an example, with a velocity of $v = 0{,}001$ m s^{-1} inside a capillary of ID $= 50$ μm, the following result is obtained:

$$R = \frac{\rho_m v d}{\eta} = 0{,}05 \ll 10 ,$$

where ρ_m is the specific mass (density) and η is the dynamic viscosity. Therefore, the probability of observing the onset of any turbulence under these conditions [11] is negligible.

D.3 Zeta Potential

The results found in the previous section made use of the approximation C.11.

$$\sinh\left(\frac{Ze\Psi}{kT}\right) \simeq \frac{ze\Psi}{kT} .$$

Moreover, a second subtle approximation was made at the boundary condition C.19, which is not in exact accordance with the measurement:

$$v_x(a) = 0 .$$

The experimental results from [1] and [12] show that the shear plane (or slipping plane) occurs many nanometers away from the dielectric surface. At the shear plane the electric potential is no longer Ψ_0, but it is lower. The potential at the shear plane is represented by $\Psi = \zeta$ (the "zeta" potential or electrokinetic potential). A more realistic model introduced by Stern, [13] which Hunter [1] called the Gouy–Chapman–Stern model, assumes that the aqueous solution contains a few layers of water molecules mixed together with some ions (that are expected to belong to the diffuse layer) that are immobilized at the surface. These water molecules are held by the permanent dipole–charge interactions, hydrogen bonding and Van der Walls forces in some cases. This is called Stern layer and is represented by the red rectangles in Fig. 2.17. Therefore, the Poisson–Boltzmann equation D.8 applies for $0 \leq r \leq a'$, but not for $a' \leq r \leq a$, see Fig. 2.17.

The exact expression of the potential $\Psi(y)$ in the range $a' \leq r \leq a$ will be not treated in this book. However, it is possible to find the expression of the electroosmotic velocity $v_x(0)$. This is very important, as the total velocity of the bands in the electromigration techniques will be the sum of the electrophoretic velocity plus the electroosmotic velocity. Moreover, knowledge of the electroosmotic phenomena allows the development and optimization of strategies to control it, such as controlling the buffer pH, buffer composition, additives used and chemical modifications that can be made on the surface wall.

Starting with equation C.17 and using $P_x = 0$, the following is produced:

$$\frac{1}{r}\frac{d}{dr}\left(r\frac{dv_x}{dr}\right) = -\frac{F_x}{\eta} , \tag{D.19}$$

where, again, F_x is the drag force that results from the action of the applied field E_x on the charges represented by the distribution $\rho(r)$, which is:

$$\frac{1}{r}\frac{d}{dr}\left(r\frac{dv_x}{dr}\right) = -\frac{E_x}{\eta}\rho(r). \tag{D.20}$$

However, from the Poisson–Boltzmann equation C.9 is it possible to obtain:

$$\rho(r) = -\epsilon \frac{1}{r} \frac{d}{dr} \left(r \frac{d\Psi}{dr} \right) . \qquad (D.21)$$

Therefore,

$$\frac{d}{dr} \left(r \frac{dv_x}{dr} \right) = \frac{\epsilon E_x}{\eta} \frac{d}{dr} \left(r \frac{d\Psi(r)}{dr} \right) . \qquad (D.22)$$

This equation can be integrated twice from $r = 0$, where $\Psi(0) = 0$ and $v_x(0) = v_{eo}$, until the shear plane where $\Psi(a - \delta) = \zeta$ and $v_x(a - \delta) = 0$. Using, for the first integration, the fact that $\left[\frac{d\Psi}{dr} \right]_{r=0}$ and $\left[\frac{dv_x}{dr} \right]_{r=0}$ are null, gives:

$$v_{eo} = -\frac{\epsilon \zeta E_x}{\eta} \qquad (D.23)$$

or

$$\zeta = -\frac{\eta v_{eo}}{\epsilon E_x} \qquad (D.24)$$

and

$$\mu_{eo} = -\frac{\epsilon \zeta}{\eta} . \qquad (D.25)$$

The variables η, v_{eo} and ϵ are experimentally measurable quantities, therefore the potential ζ may be determined. Equation D.24 is known as the Smoluchowski equation.

The potential ζ is widely measured and used in the literature, especially in the field of physics and chemistry of interfaces,[1] as many electrokinetic phenomena are directly dependent on the ζ potential.

This appendix has presented a succinct description of the concepts pertaining to the electric double-layers and both the Gouy–Chapman and Gouy–Chapman–Stern models. The resulting mathematical expressions of the electroosmotic velocity and the velocity profiles were given for the general case, which is the concomitant action of both electroosmosis and Poiseuille flows, the latter occurring due to pressure gradients. The shown graphs are important for understanding how the strategy called Poiseuille counter-flow has the potential to mitigate band broadening caused by thermal gradients that are created when the ESTs are operated under strong electric fields. The operation under this strong field strengths improves remarkably the performance parameters as summarized in Tables 5.4 and 5.5.

References

1 Hunter, R.J. (1981). *Zeta Potential in Colloid Science Principles and Applications*. London: Academic Press.

2 Smoluchowski, M. (1921). *Handbuch der Electrizität und des Magnetismus*, 336. Leipzig: Barth.

3 Gouy, M. (1910). *Journal de Physique Théorique et Appliqué* 9: 457–468.

4 Chapman, D.L. (1913). *Philosophical Magazine* 25: 475–481.

5 Debye, P. and Hückel, E. (1923). *Physikalische Zeitschrift* 24: 185–206.

6 Elton, G.A.H. (1950). *Proceedings of the Royal Society of London Series A: Mathematical and Physical Sciences* 198: 581–589.

7 Burgreen, D., Nakache, F.R. (1964). *The Journal of Physical Chemistry* 68: 1084–1091.

8 Rice, C.L. and Whitehead, R. (1965). *The Journal of Physical Chemistry* 69: 4017–4024.

9 Levine, S., Marriott, J.R., Neale, G., and Epstein, N. (1975). *Journal of Colloid and Interface Science* 52: 136–149.

10 Gradshteyn, I.S. and Ryzhik, I.M. (1965). *Table of Integrals Series and Products*, 951. New York: Academic Press.

11 Landau, L.D. and Lifshitz, E.M. (1959). *Fluid Mechanics*, 62 and 103. London: Pergamon Press.

12 Schwer, C. and Kenndler, E. (1991). *Analytical Chemistry* 63: 1801–1807.

13 Stern, O.Z. (1924). *Zeitschrift für Elektrochemie* 30: 508–516.

Appendix E

Molecular Diffusion

E.1 The Diffusion Equation

This appendix discusses the diffusion of atoms, molecules, ions, and particles (the analytes) within separation media that have low Reynolds numbers. In liquids, which are the most common separation media in ESTs, a low Reynolds number means that no convection (warmer parts of the liquid move up and cooler parts move down) occurs. If the liquid flows, then it flows in a laminar way (without turbulence). It is considered that the media are isotropic and homogeneous, and that the diffusing analytes are present at low concentrations, such that the physical properties of the medium remain unchanged. In this case the *molecular diffusion*, also called self-diffusion,[1] is the predominant phenomenon. It originates from Brownian motion, which is the random motion of the analytes caused by collisions with the fast-moving molecules of the liquid.

Molecular diffusion studies deal with the time evolution of mass concentration distributions or amount concentration (molar) distributions of the analytes. It is represented by $c(\vec{r}, t)$, where the units of c can be mass/volume, mol/volume, or any other unit of concentration. The concentration c at point $\vec{r}$ and time t is calculated by dividing the number of moles (n) that exist within the small volume V_{small} (at time t and centered at $\vec{r}$) by V_{small}. However, the volume V_{small} must be large enough to avoid the random fluctuations of n caused by thermal energy and, at the same time, it is necessary that $V_{small} \ll V_{total}$, where V_{total} is the total volume of the separation phase (liquid or gel). If at any time t the distribution $c(\vec{r}, t)$ is normalized to one (its integral over the entire space is equal to one) then it is a *probability density function*. This is a non-negative function that is related to the probability (P) of finding a single analyte within the infinitesimal volume $\delta V = \delta x \, \delta y \, \delta z$ at time t, i.e. $P \propto c(\vec{r}, t) \, \delta x \, \delta y \, \delta z$.

Open and Toroidal Electrophoresis: Ultra-High Separation Efficiencies in Capillaries, Microchips, and Slabs, First Edition. Tarso B. Ledur Kist.
© 2020 John Wiley & Sons Ltd. Published 2020 by John Wiley & Sons Ltd.

If there are no external forces acting on the diffusing analytes then the points of the distribution c within the media that have higher values become less concentrated (c decreases) and the points with lower values become more concentrated (c increases). This happens due to the random nature of Brownian motion. After some time, the concentration c tends asymptotically to a fixed value at all positions within the volume V_{total} if the inlet and outlets are blocked, if not the entities (ions or molecules) will also diffuse into the reservoirs. In thermodynamics this is called the tendency of *entropy* maximization.

The current density of molecules ($\vec{J}$) is an important parameter that is defined as the net flow of molecules that cross a given area of a flat plane per unit time. If c is constant at all points within the volume V_{total} then $\vec{J}$ is expected to be zero. However, if c is not constant in a given direction, for instance along x, then $J_x = \partial J/\partial x$ must be proportional to the gradient of c in the x direction. Mathematically, it can be expressed as:

$$J_x \propto -\frac{\partial c}{\partial x}.$$

This expression can be deduced using the kinetic theory of liquids and gases.[2] In three dimensions it becomes:

$$\vec{J}(\vec{r}, t) = -\overleftrightarrow{a} \cdot \vec{\nabla} c(\vec{r}, t), \tag{E.1}$$

where $\vec{\nabla}$ is the nabla operator. This is Fick's first law,[1] where $\overleftrightarrow{a}$ is the diffusion tensor of order two and depends on the physical characteristics of the medium. Considering that the diffusing species (analytes) are stable (there are no sinks and sources), then c also satisfies the continuity equation:

$$\frac{\partial c}{\partial t} = -\vec{\nabla} \cdot \vec{J}. \tag{E.2}$$

Combining equations E.1 and E.2 produces the following:

$$\frac{\partial c}{\partial t} = \vec{\nabla} \cdot (\overleftrightarrow{a} \cdot \vec{\nabla} c). \tag{E.3}$$

If the separation medium is isotropic then $a_{ij} = \delta_{ij} D$, and if D is both position and time independent then:

$$\frac{\partial c}{\partial t} = D \nabla^2 c \quad \text{or}$$

$$= D \left(\frac{\partial^2 c}{\partial x^2} + \frac{\partial^2 c}{\partial y^2} + \frac{\partial^2 c}{\partial z^2} \right), \tag{E.4}$$

where D is the molecular diffusion coefficient. This is Fick's second law.[1] It is identical to the heat diffusion equation (without a heat source), with the exception of D, which must be replaced by the thermal conductivity of the medium (λ). Many other phenomena are also described by this equation, such as the diffusion of photons in an non-absorbent, dispersive media.

The majority of the diffusion phenomena that occur in ESTs follow, in good approximation, this linear partial differential equation.(E.4) Moreover, most separation modes are uni-dimensional, which simplifies equation E.4 even further:

$$\frac{\partial c}{\partial t} = D\frac{\partial^2 c}{\partial x^2}. \tag{E.5}$$

The solutions of equation E.5 are surfaces over a plane defined by the independent variables x and t, as shown in Fig. 2.7.

E.2 The Propagator

In most applications the initial distributions $c(\vec{r}, t)$ of the analytes at time t are known. In the ESTs this corresponds to the initial spatial profile of the injected sample plug. The goal of this section is to find the shape of this distribution at any time t' that is larger than t. Throughout this section the center of mass of the distributions will be considered to be at rest all the time. A powerful way of treating this problem is to use *Green's function*, which Butkov [2] has called "Green's function for the initial condition". It is widely applied in quantum mechanics,[3,4] where it is also called the "retarded Green's function" or simply the "propagator".

If the partial differential equation E.4 is linear and of the first order in time, then it makes sense to suppose that a function (G) that satisfies the following equation exists:

$$c(\vec{r}\,', t') = \int G(\vec{r}\,', t'; \vec{r}, t)c(\vec{r}, t)\mathrm{d}^3 r, \tag{E.6}$$

where $G(\vec{r}\,', t'; \vec{r}, t)$ is Green's function. Moreover, there is a component of Green's function (G^+) that propagates $c(\vec{r}, t)$ in the positive direction of time (forward). However if the distribution of an analyte is known at time t' and the distribution at a previous time needs to be calculated, then the component G^-, which propagates $c(\vec{r}, t)$ backwards in time, must be used. G^+ is more interesting than G^- because it allows the behavior of the distributions to be predicted. Therefore, equation E.6 can be re-written in terms of the component G^+:

$$\theta(t' - t)c(\vec{r}\,', t') = \int G^+(\vec{r}\,', t'; \vec{r}, t)c(\vec{r}, t)\mathrm{d}^3 r, \tag{E.7}$$

where $\theta(t' - t)$ is the Heaviside function and $G^+(\vec{r}', t'; \vec{r}, t)$ is the propagator of differential equation E.4. As $c(\vec{r}, t)$ and $c(\vec{r}', t')$ must satisfy equation E.4, applying the operator $\left(\frac{\partial}{\partial t'} - D\nabla'^2\right)$ to equation E.7 will produce:

$$\left(\frac{\partial}{\partial t'} - D\nabla'^2\right)G^+(\vec{r}\,', t'; \vec{r}, t) = \delta^3(\vec{r}\,' - \vec{r})\delta(t' - t), \tag{E.8}$$

where $\delta(x)$ is the Dirac delta function (distribution). Equation E.8 is known as the partial differential equation that must be satisfied by the forward propagator G^+. When deriving this equation the separation media are assumed to be homogeneous and isotropic in space and time. Using the properties of the Fourier transform it is easy to obtain G_F^+ from equation E.8, and applying the inverse Fourier transform to $G^+(\vec{r}\,',t';\vec{r},t)$ generates the following expression:

$$G^+(\vec{r}\,',t';\vec{r},t) = \left(\frac{1}{4\pi D(t'-t)}\right)^{3/2}\left\{\exp\left[\frac{-|\vec{r}\,'-\vec{r}|^2}{4D(t'-t)}\right]\right\}\theta(t'-t). \qquad \text{(E.9)}$$

This is the forward propagator of the molecular diffusion equation E.4 (replace r by x and 3/2 by 1/2 in the one-dimensional case). From now on any $c(\vec{r}\,',t')$ can be predicted using equation E.6 as long as $c(\vec{r},t)$ (the initial condition) is known.

E.3 Application of Propagators to Bands at Rest

The separation paths used in ESTs typically range from ~1 cm to ~1 m in length. These paths can be filled with gels, polymer solutions, or liquids. When a liquid is used, the internal dimensions of the separation paths are very small (in the 20 to 100 μm range). This gives low Reynolds numbers (usually lower than ten) that prevent convection. This means that molecular diffusion (self-diffusion) can be the only band broadening mechanism present if some additional measures are taken. Obviously, many other band broadening mechanisms are possible [5,6].

Many sample injection procedures exist. However, there are only two that produce a sample plug with the rectangular shape shown in Figure E.1. These methods are: electrokinetic sample injection, which is used in the microchip and capillary platform, and micropipetting samples into the wells of gel slabs (e.g., polyacrylamide and agarose). Therefore at time $t = 0$, $c(\vec{r},0) = c_0$ if $-b < x < b$, but $c(\vec{r},0) = 0$ if $x < -b$ and $x > b$.

Therefore at time t, which can be considered as $t = 0$, the concentration of molecules along the path will be $c(\vec{r},0)$, as shown in Figure E.1.

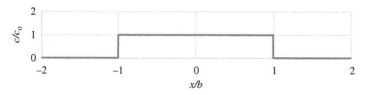

Figure E.1 Spatial concentration profile of an analytes at $t = 0$. This is observed when the electrokinetic injection mode is used (in the microchip or capillary platform) or when the sample is dispensed into a well of a polyacrylamide or agarose gel slab using a micropipette. The sample plug is considered centered at $x = 0$ and has a length of $2b$.

Note that concentration profiles are constant in the y and z directions. Therefore, the concentration profile along x is the only thing that changes over time, which reduces the equations to the following one-dimensional definite integral:

$$c(x',t') = \int_{-\infty}^{\infty} G^+(x',t';x,0)c(x,0)dx \tag{E.10}$$

with the following initial condition:

$$c(x,0) = \begin{cases} c_0 & \text{if} -b < x < b \\ 0 & \text{if } x < -b \text{ or } x > b. \end{cases} \tag{E.11}$$

This is very easy to solve. Its solution is given by:

$$c(x',t') = \frac{c_0}{2}\left\{ \text{erf}\left(\frac{b-x'}{2\sqrt{Dt'}}\right) + \text{erf}\left(\frac{b+x'}{2\sqrt{Dt'}}\right) \right\}, \tag{E.12}$$

where the erf function is defined as [7]:

$$\text{erf}(z) = \frac{2}{\sqrt{\pi}}\int_0^z e^{-u^2}du.$$

The solution given by equation E.12 is valid for any case where $t' > 0$. From now on, (x, t) will be used instead of (x', t') to simplify the notation. Equation E.12 can be expanded into a MacLaurin series and produces the following:

$$c(x,t) \simeq \frac{bc_0}{\sqrt{\pi Dt}}e^{-x^2/4Dt}. \tag{E.13}$$

This approximation is good if $b/\sqrt{Dt} \ll 1$. In this case, a sample plug that is rectangular at $t = 0$ evolves into a Gaussian (normal) distribution along x if it is left to diffuse for long enough (the time necessary depends on b and D). Moreover, Einstein's relationship $\sigma^2 = 2Dt$ becomes very explicit from equation E.13, i.e. variance grows as $2Dt$ and the distribution (band) spreads indefinitely in time.

When these distributions are measured by a detection system using a scanner [8], the detected signal $s(x, t)$ is a convolution of equation E.12 with the spatial filtering function $g(x)$ (which depends on the slit width) and it is multiplied by the sensibility factor ϕ (which depends on the detector used and the sensibility of the detector). The function $g(x)$ exists because it is not possible to detect the analytes at a single cross sectional plane of the separation medium for many reasons. Firstly, because it is almost impossible to find the center of mass of a single ion or molecule located at such an hypothetical plane at time of detection τ. Secondly, the sensing parts of the detection systems always require a certain spatial extension (related to the size of the electrodes in electrochemistry, width of the pixels within a CCD, width of the slit of a PMT or APD, to cite a few). The function $g(x)$ is always positive and is usually symmetrical with respect to the center of the length l_{det}. Therefore, the experimentally measured signal $s(x, t)$ is always related to $c(x, t)$ by the following convolution integral:

$$s(x, t) = \phi \frac{bc_0}{\sqrt{\pi Dt}} \int_{-\infty}^{\infty} e^{-(x-\xi)^2/4Dt} g(\xi) d\xi. \tag{E.14}$$

If $l_{\text{det}} \ll \sqrt{Dt}$ then it is possible to show that $s(x, t) \simeq \phi c(x, t)$. If this condition is not fulfilled the convolution integral must be applied to $c(x, t)$ in order to get $s(x, t)$ and the deconvolution integral must be applied to $s(x, t)$ to get $c(x, t)$.

E.4 Application of Propagators to Bands in Movement

In ESTs the analytes move due to the application of the electric field E. This movement comes from electroosmosis and/or the movement of the analytes with respect to the liquid of the separation medium. This movement of the analytes with respect to the separation medium occurs due to one (or a combination) of the following two phenomena: (1) Movement is caused by electrophoresis and given by $v = \mu_e E$, where μ_e is the electrophoretic mobility. Note that only charged analytes (ions, charged molecules, or charged particles) are subjected to this electrophoresis. (2) Charged additives interact with the neutral analytes in the following way: analyte + additive $\leftrightharpoons$ complex (or aggregate). The analytes are carried by the additives as part of the complex for many short time periods, and are left motionless in the interims (when separated). This net average velocity is given by $v = \mu_{el} E$, where μ_{el} is called electrical mobility.

If electroosmosis is present or unknown then the term μ, called apparent mobility (mobility with respect to the platform wall), is used and it does not specify which of the cited phenomena is responsible for the movement. If a given analyte moves with a constant velocity of $v = \mu E$, then the time evolution of this molecular distribution can still be represented by equation E.12. The only thing that needs to be done is to change the referential frame from x' to $x - \mu E t$ [9]. Therefore, the general solution of the time evolution of a molecular distribution that has the rectangular shape shown by equation E.11 and Figure E.1 at $t = 0$ is given by:

$$c(x, t) = \frac{c_0}{2} \left\{ \text{erf}\left(\frac{b + \mu E t - x}{2\sqrt{Dt}} \right) + \text{erf}\left(\frac{b - \mu E t + x}{2\sqrt{Dt}} \right) \right\}. \tag{E.15}$$

Again, if $b/\sqrt{Dt} \ll 1$ then:

$$c(x, t) \simeq \frac{bc_0}{\sqrt{\pi Dt}} e^{-(x-\mu E t)^2/4Dt}. \tag{E.16}$$

This is the time evolution of the bands which are subjected to molecular diffusion (represented by D) and an electrodriven movement (represented by $v = \mu E$).

E.5 Bands and Peaks

The definitions of bands and peaks have already been discussed in Section 2.6 and were made based on Figure 2.7. It is now possible to make a mathematical distinction between bands and peaks using equations E.15 and E.16.

The bands are detected at time τ along the separation path using a spatial scanner (in the microchip and capillary platforms) or a photographic camera within a transilluminator (in the slab platform). Therefore, fixing the time (of detection) as τ in equations E.15 and E.16 produces what follows:

$$s(x, \tau) = \sum_{i}^{n} \phi_i \frac{c_o}{2} \left\{ \mathrm{erf}\left(\frac{b + \mu_i E \tau - x}{2\sqrt{D_i \tau}} \right) + \mathrm{erf}\left(\frac{b - \mu_i E \tau + x}{2\sqrt{D_i \tau}} \right) \right\}, \quad (E.17)$$

where n is the total number of analytes. This is the spatial profile of the bands along the separation path at time τ. If they are all separated from each other then the total number of separated bands will coincide with the total number of analytes.

Let us suppose that the detection length (l_{det}) of the spatial scanner is such that $b/\sqrt{Dt} \ll 1$. In this case the signal produced by the detector of the n bands (analytes) is given by:

$$s(x, \tau) \simeq \sum_{i}^{n} \frac{\phi_i bc_o}{\sqrt{\pi D_i \tau}} e^{-(x - \mu_i E \tau)^2 / 4 D_i \tau}. \quad (E.18)$$

Note that the bands are symmetric (Gaussian), as x is the only independent variable (τ is a constant).

However, if a detector is fixed at position $x = l$ to detect the passage of the bands at this position along the course of time, then the recorded signal will be different and given by:

$$s(l, t) = \sum_{i}^{n} \phi_i \frac{c_o}{2} \left\{ \mathrm{erf}\left(\frac{b + \mu_i E t - l}{2\sqrt{D_i t}} \right) + \mathrm{erf}\left(\frac{b - \mu_i E t + l}{2\sqrt{D_i t}} \right) \right\}. \quad (E.19)$$

Again, if $b/\sqrt{Dt} \ll 1$, then

$$s(l, t) \simeq \sum_{i}^{n} \frac{\phi_i bc_o}{\sqrt{\pi D_i t}} e^{-(l - \mu_i E t)^2 / 4 D_i t}. \quad (E.20)$$

Equations E.19 and E.20 give the shape of the peaks in the electropherogram. Note that equation E.20 is neither Gaussian nor symmetric as the independent variable is now t and it appears in three places of the right side of equations E.20.

References

1 Shewmon, P.G. (1963). *Diffusion in Solids*, 2–5. New York: McGraw-Hill.
2 Butkov, E. (1968). *Mathematical Physics*. Boston, MA: Addison-Wesley.
3 Schiff, L.I. (1968). *Quantum Mechanics*, Chapter 9. Tokyo: McGraw-Hill.
4 Bjorken, J.D. and Drell, S.D. (1964). *Relativistic Quantum Mechanics*, Chapter 6. New York: McGraw Hill.
5 Huang, X., Coleman, W.F., and Zare, R.N. (1989). *Journal of Chromatography* 480: 95–110.
6 Roberts, G.O., Rhodes, P.H., and Snyder, R.S. (1989). *Journal of Chromatography* 480: 35–67.
7 Abramowitz, M. and Stegun, I.A. (1964). *Handbook of Mathematical Functions: With Formulas, Graphs, and Mathematical Tables*. Washington, DC: National Bureau of Standards.
8 Beale, S.C. and Sudmeier, S.J. (1995). *Analytical Chemistry* 67: 3367–3371.
9 Schoffen, J.R., Mandaji, M., Termignoni, C. et al. (2002). *Electrophoresis* 23: 2704–2709.

Appendix F

Poiseuille Counter-flow

F.1 Introduction

The application of high voltages and strong electric fields is important for all ESTs as they optimize the performance indicators in the ways summarized in Tables 5.2–5.5. The performance indicators shown in Table 5.4 are easier to understand because L is eliminated when possible. Applied electric field strength is an important parameter for some performance indicators, while the applied electrostatic potential difference (voltage) is what matters for others. Simply put, the performance indicators are affected by V and E in the following ways:

Plate number	N $\propto$	V
Plate double-rate	$\mathcal{N} = N/t^2$ $\propto$	E^3
Plate height	H $\propto$	E^{-1}
Resolution	R $\propto$	$V^{1/2}$
Resolution rate	$R_t = R/t$ $\propto$	$E^{3/2}$
Band capacity	B $\propto$	$V^{1/2}$
Band rate	$B_t = B/t$ $\propto$	$E^{3/2}$
Peak capacity	P $\propto$	$V^{1/2}$
Peak rate	$P_t = P/t$ $\propto$	$E^{3/2}$

Note: Plate height (H) is the only performance indicator for which the smallest values are the most desired.

All performance indicators can be optimized by either maximizing the applied voltage or the electric field strength (depending on the performance indicator of interest). When high voltages are applied then high temperature gradients can be avoided by using a long separation medium (large L). However, when strong electric fields are applied (to optimize $\mathcal{N}$, H, R_t, B_t, and P_t) then high radial temperature gradients are created. These temperature gradients, in

Open and Toroidal Electrophoresis: Ultra-High Separation Efficiencies in Capillaries, Microchips, and Slabs, First Edition. Tarso B. Ledur Kist.
© 2020 John Wiley & Sons Ltd. Published 2020 by John Wiley & Sons Ltd.

Figure F.1 Illustration of the perspective view and frontal view of an orthogonal cut of an example of a conduit with an arbitrary cross sectional geometry. The black curve shows the separation medium/wall interface. When the cooling design is made in such a way to create a constant temperature over all points of this black curve, then the Poiseuille counter-flow may succeed in canceling out the thermally induce band dispersion caused by the application of strong electric fields.

turn, create radial electrophoretic velocity gradients, which produce an extra band dispersion mechanism quantified by the dispersion coefficient denoted by D_{temp} (see Section 2.13). In this case the variance will grow in time as $\sigma^2(t) = \sigma^2(0) + 2Dt + 2D_{temp}t$ providing no other band broadening mechanisms are present. This is an important band broadening mechanism that limits the electric field strengths that can be applied. Note that the dependency of mobility on temperature is discussed in Section C.2.

The following subsections will show that if the separation track of ESTs are effectively symmetrically cooled, that is, if the temperature is consistent at every point of the conduit's inner surface, then the band dispersion represented by D_{temp} can be partially canceled out by applying a Poiseuille counter-flow. This in turn will allow the application of stronger electric fields, which will improve all performance indicators. Moreover it will be shown that this is true for the general case, meaning that it can be applied to cylindrical capillaries, square capillaries, rectangular microchannels, and any microconduit with an arbitrary geometry. The only conditions are that the conduit's cross sectional geometry must remain the same along the whole extension of the separation path, and that the cooling geometry must allow the temperature to be constant along the whole liquid phase/wall interface (conduit's inner surface), which is represented by the black curve in Figure F.1.

F.2 Velocity Level Contours

The Navier–Stokes equations are a system of non-linear partial differential equations that describe both the field of velocities (v) and the pressure of a viscous fluid. In Cartesian coordinates they are written in the following way:

$$\rho\left(\frac{\partial v_x}{\partial t} + v_x\frac{\partial v_x}{\partial x} + v_y\frac{\partial v_x}{\partial y} + v_z\frac{\partial v_x}{\partial z}\right) = F_x + \frac{\partial P}{\partial x} + \eta\left(\frac{\partial^2 v_x}{\partial x^2} + \frac{\partial^2 v_x}{\partial y^2} + \frac{\partial^2 v_x}{\partial z^2}\right),$$

$$\rho\left(\frac{\partial v_y}{\partial t} + v_x\frac{\partial v_y}{\partial x} + v_y\frac{\partial v_y}{\partial y} + v_z\frac{\partial v_y}{\partial z}\right) = F_y + \frac{\partial P}{\partial y} + \eta\left(\frac{\partial^2 v_y}{\partial x^2} + \frac{\partial^2 v_y}{\partial y^2} + \frac{\partial^2 v_y}{\partial z^2}\right),$$

$$\rho\left(\frac{\partial v_z}{\partial t} + v_x\frac{\partial v_z}{\partial x} + v_y\frac{\partial v_z}{\partial y} + v_z\frac{\partial v_z}{\partial z}\right) = F_z + \frac{\partial P}{\partial z} + \eta\left(\frac{\partial^2 v_z}{\partial x^2} + \frac{\partial^2 v_z}{\partial y^2} + \frac{\partial^2 v_z}{\partial z^2}\right),$$

$$\text{and} \quad \frac{\partial v_x}{\partial x} + \frac{\partial v_y}{\partial y} + \frac{\partial v_z}{\partial z} = 0, \tag{F.1}$$

where ρ is the density of the liquid, F_x, F_y, and F_z are the components of a force field, P is the applied pressure, and η is the dynamic viscosity of the liquid. For unidirectional, laminar, and fully developed flows (i.e. neglecting the entrance and exit regions) the following is obtained:

$$\rho\left(\frac{\partial v_x}{\partial t}\right) = F_x + \frac{\partial p}{\partial x} + \eta\left(\frac{\partial^2 v_x}{\partial y^2} + \frac{\partial^2 v_x}{\partial z^2}\right). \tag{F.2}$$

In ESTs the term F_x is responsible for the EOF and, for now, it can be neglected to simplify the equations. Moreover, in the great majority of ESTs the steady state is the most important regime (the transient regimes are discussed in Sections C.3.1 and C.3.2). Therefore, considering $F_x = 0$ and $\partial v_x/\partial t = 0$ the following is obtained:

$$\frac{\partial^2 v_x}{\partial y^2} + \frac{\partial^2 v_x}{\partial z^2} = -\frac{\Delta P}{\eta L}. \tag{F.3}$$

It was assumed that $\partial P/\partial x = \Delta P/L$, where ΔP is the pressure difference applied between reservoirs separated by a distance of L.

Equation F.3 is known as Poisson's equation with a non-homogeneous term. Its solutions for cylindrical conduits are given in Section 2.8.3 and for square and rectangular conduits in Section 2.8.4. In these subsections the velocity profiles are shown as contour levels. It is important to keep in mind that equation F.3 can be used to find these velocity profiles for any arbitrary geometry by using, for instance, numerical integration using finite elements. The only condition (besides the ones cited above) is that the conduit geometry must be identical along its whole extension. Figure F.1 shows such a conduit with an arbitrary geometry.

F.3 Temperature Level Contours

The level contours of temperature profiles inside the cross sectional area of the separation medium are given by a heat diffusion equation containing a non-homogeneous scalar term called "heat source". This term represents the Joule effect and is discussed in Section 2.12 in the cylindrical coordinates. Here we will discuss it in the Cartesian coordinates, which are the most commonly used coordinates in the general case (i.e. for conduits with an arbitrary geometry as shown by Figure F.1). By taking equation 2.32 and making the coordinate changes $y = r\cos(\theta)$ and $z = r\sin(\theta)$ the following is produced:

$$\frac{\partial^2 T}{\partial y^2} + \frac{\partial^2 T}{\partial z^2} = -\frac{p}{\lambda}, \tag{F.4}$$

where p is the heat generated per unit volume and unit time due to the Joule effect, and λ is the thermal conductivity coefficient of the separation medium.

F.4 Equalizing v_{max} and Δv_e

Surprisingly, the solutions of equations F.3 and F.4 are identical in the general case, i.e. for conduits with a cylindrical, square, rectangular, or any arbitrary geometry (as long as the border conditions used are the same). In other words, the level contours of the velocity profiles and the level contours of the temperature profiles (and electrophoretic velocities) will be identical if temperature is constant over the whole inner surface (separation medium/wall interface) of the conduit shown in Figure F.1. This surface is represented by the black curve in Figure F.1. It is assumed that the pressure driven velocity of the separation medium is zero at this liquid/wall interface, which is the case in all ESTs.

From a mathematical point of view, the heat diffusion equation containing a scalar term (heat source) is identical to the Navier–Stokes equation for unidirectional, laminar, fully developed, non-compressible viscous fluids in the steady state. This is true for the general case. The solutions of these differential equations will be identical if v and T are constant at the boundary formed by the inner wall of the conduit (capillary, microchannel, or any other conduit). The boundary condition for the Navier–Stokes equation is that velocity is zero, which is observed across all ESTs. Therefore, temperature must also be constant at all points of this boundary (surface). If this condition is fulfilled then the effect of band broadening caused by temperature gradients can be exactly canceled out by using a fine-tuned pressure induced counter-flow. In this case, the electromigrations conducted under strong electric fields do not present the undesirable effect of thermal induced band broadening. However, this only applies to a certain extent because the dynamic viscosity of the separation medium is not constant, as is considered in equation F.3.

In the case of cylindrical conduits it is easy to calculate the pressure difference (ΔP) that must be applied between the ends of a capillary (of length L) to cancel the band broadening effect. In this case v_{max}, given by equation 2.14, must exactly cancel the Δv_e (electrophoretic velocity difference) created by the temperature difference ΔT. By combining equations 2.31 and 2.33, the following is produced:

$$\Delta P = \frac{k\eta ELp}{\lambda}\mu_e(T_a),$$ (F.5)

where ΔP is the pressure difference that must be applied to a capillary (of length L) to cancel the band dispersion mechanism caused by the temperature gradient, η is the dynamic viscosity of the liquid, p is the heat generated by the Joule effect ($p = Vi/(\pi a^2 L)$) per unit volume and unit time, λ is the thermal conductivity coefficient (e.g., in W m^{-1} K^{-1}) of the separation medium (liquid), k is the dependence of the mobility on temperature (k has a typical value of $\simeq 0.025/°C$), and $\mu_e(T_a)$ is the electrophoretic mobility of the analyte at the temperature T_a (the temperature at the liquid/wall interface). Equation F.5 can be rewritten using the operational parameters, which produces:

$$\Delta P = \frac{k\eta V^2 i}{\lambda \pi a^2 L}\mu_e(T_a),$$ (F.6)

where V is the applied voltage, i is the total current observed, and a is the inner radius of the capillary.

When the gravimetric method is used to drive the Poiseuille counter-flow the following level difference (ΔH) between reservoirs must be used:

$$\Delta H = \frac{k\eta V^2 i}{\rho g \lambda \pi a^2 L}\mu_e(T_a),$$ (F.7)

where ρ is the density of the liquid and g is the local gravity acceleration.

In the case of square or rectangular microchannels, which are commonly used in microchip electrophoresis, the following is obtained:

$$\Delta P = \frac{k\eta V^2 i}{\lambda whL}\mu_e(T_a)$$ (F.8)

and

$$\Delta H = \frac{k\eta V^2 i}{\rho g \lambda whL}\mu_e(T_a),$$ (F.9)

where w and h are, respectively, the width and height of the microchannel.

For the general case of a conduit with an arbitrary geometry, the following is obtained by using logical arguments and equations F.6, F.7, F.8, and F.9:

$$\Delta P = \frac{k\eta V^2 i}{\lambda AL}\mu_e(T_a)$$ (F.10)

and

$$\Delta H = \frac{k\eta V^2 i}{\rho g \lambda AL}\mu_e(T_a),$$ (F.11)

where A is the cross sectional area of a microconduit with an arbitrary geometry represented by Figure F.1.

The presence of EOF does not change the results of this appendix. This can be seen by analyzing the results of Section 2.9, especially Figure 2.18. The presence of EOF can be interpreted merely as a change in the referential frame of equations F.1 and F.4. However, this is only valid if the thickness of the diffuse layer ($1/\kappa$; see Figure 2.17) is much smaller than the smallest radius of curvature of the inner surface of the conduit.

A result similar to this Appendix was obtained for the specific case of capillary electrophoresis by Gobie and Ivory [1].

Reference

1 Gobie, W.A. and Ivory, C.F. (1990). *Journal of Chromatography* 516: 191–210.

Appendix G

Cyclic On-column Band Compression

G.1   Introduction

In this appendix the effects that result from applying cyclic band compression events to the toroidal layout will be presented. Band compression is an on-column technique used to compress the widths of bands during the runs [1,2], which in turn will improve band capacity and peak capacity. This desired result is achieved by using, for instance, an independent thermal zone with temperature programming. This has been demonstrated for the open capillary free-solution electrophoresis [1,2]. In addition it can be used in other platforms (microchip and slab) and separation modes, including: affinity electrophoresis (AE), end-labeled free-solution electrophoresis (ELFSE), micellar electrokinetic chomatography (MEKC), and micro-emulsion electrokinetic chomatography (MEEKC). However, its integration into electrochromatogrphy (EC), isoelectric focusing (IEF), and isotchophoresis (ITP) would require further work to determine its feasibility and usability.

Stacking is a different procedure that involves compressing the injected sample plug at the beginning of runs. It is used to improve detection limits and separation efficiencies [3–9]. The differences between band compression and stacking are detailed in Section 2.14 and illustrated by Figures 2.20 and 2.21.

Analytes run in a cyclic manner in toroidal ESTs. This is a very appealing property as repeated on-column band compression events can have remarkable effects on the separation process. To facilitate the calculation of the time evolution of the variance ($\sigma^2(t)$) and band spacing ($\Delta x(t)$) in the next sections, a model with the following properties will be adopted:

(1) Analytes run in a clockwise ($\circlearrowright$) manner and their concentration is much lower than that of the BGE. Therefore, band compression will not induce electrokinetic dispersion.

(2) The thermal zone is much shorter than the total length of the toroid. This simplifies the calculations but compromises band capacity.

Open and Toroidal Electrophoresis: Ultra-High Separation Efficiencies in Capillaries, Microchips, and Slabs, First Edition. Tarso B. Ledur Kist.
© 2020 John Wiley & Sons Ltd. Published 2020 by John Wiley & Sons Ltd.

(3) The ratio of the velocity of the analytes inside the thermal zone to the velocity of the analytes outside the thermal zone is given by α. This is the compression factor that can be achieved by (at least) the two methods described in Section 2.14. Once the analytes are inside the compression zone the conditions of the thermal zone are reversed back to the normal running conditions.

(4) A torus with four microholes (four reservoirs) and a total length of $2L$ is considered.

(5) For simplicity, two oppositely positioned fixed detectors will be considered, one at $x = 0$ (the injection point) and the other at $x = L$.

(6) The two thermal zones are also positioned opposite each other and may occupy any position between the following two extremes: both being located just after each detector or just before each detector.

(7) The time for an analyte to run a distance of L is given by τ.

(8) The analytes spend the time interval of $\simeq \delta\tau$ in the thermal zone. The time interval $\delta\tau$ is considered much smaller than τ ($\delta\tau \ll \tau$).

Some of the above assumptions are not very realistic, e.g., $\delta\tau \ll \tau$. However, they allow the determination of mathematical expressions for $\sigma^2(t)$ and $\Delta x(t)$, without the need to use numerical calculus. These mathematical expressions are calculated in detail in Section G.10.

G.2 Effect of Cyclic Band Compression Events on Variance

When band compression events are applied in a cyclic manner they compress both the initial variance (in a repeated manner) and the variance that molecular diffusion continuously adds to the total variance. The contribution of the initial variance is quickly extinguished within a view cycles and what is left is a steady state where an equilibrium between the molecular diffusion and cyclic compression is observed. As shown in Section G.10, other time dependent dispersion mechanisms may be present and can be added to molecular diffusion and the effect is basically the same.

Figure G.1 shows how the variance changes over time when one cyclic band compression event is performed before each detection (each detection is represented by n in the graph). Note that two detections are performed per turn and therefore two compression events occur per turn. The graph shows three distinct initial variances: $\sigma^2(0) = 0$, $\sigma^2(0)/2D\tau = 4$, and $\sigma^2(0)/2D\tau = 9$. Two compression factors were used: $\alpha = 0.8$ and $\alpha = 0.5$ (the latter was already demonstrated) [20,21]. Note that the variance reaches a steady state after four round trips (following the $n = 8$th detection) for $\alpha = 0.5$ and after eleven round trips (following the $n = 22$nd detection) for $\alpha = 0.8$.

G.3 Number of Theoretical Plates

The number of plates is defined as $N = L_{\text{tot}}/\sigma^2$. Assuming that $\sigma^2(0) \ll 2D\tau$ and using the values of σ^2 given by equations G.13 and G.14, calculated in Section G.10, the following is obtained:

$$N = \frac{n^2}{f_n} \frac{\mu E L}{2D} = \frac{n^2}{f_n} \frac{\mu V}{2D}, \qquad \text{where} \tag{G.1}$$

$$\frac{\alpha^2(1 - \alpha^{2n})}{1 - \alpha^2} \leq f_n \leq \frac{1 - \alpha^{2n}}{1 - \alpha^2}. \tag{G.2}$$

The effect of the two oppositely positioned compression zones is expressed by f_n and ranges from a minimum to a maximum value, depending where they are located. If they are located just before each detector then they produce the minimum value given by the left hand side expression of equation G.2 and if they are located just after each detector they produce the maximum value given by the right hand side expression of equation G.2.

From Table G.1 we can see that the term f_n goes quickly to a small constant value. Therefore, the number of plates increases as n^2. Moreover, if $\alpha \to 1$, which means no band compression events are applied, then the term f_n goes to n and the resulting expression is identical to equation 4.8. Mathematically, $\lim_{\alpha \to 1} f_n = n$ and, consequently, $n^2/f_n = n$ in this limit. In addition if $n = 1$, which is the case in open electrophoresis, then equation 3.2 (which represents the number of plates of open ESTs) is recovered. All this shows the consistency of equation G.1.

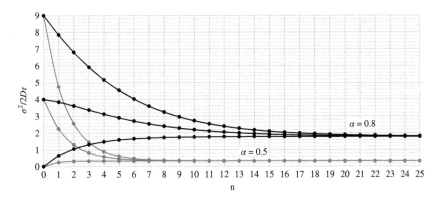

Figure G.1 The effect of cyclic band compression on the variance of the bands when a toroidal layout (closed loop path) is used. Three initial variances, related to the injected sample plug length, where tested: $\sigma(0)/2D\tau = 0, 4$, and 9. Two compression factors were used: $\alpha = 0.5$ and 0.8.

G.4 Number of Theoretical Plates per Unit Time

It is interesting to compare the number of plates delivered per unit time by different layouts and methods. This is obtained by dividing equation G.1 by $n\tau$, which is the run time of the nth detection. Assuming that the initial variance is negligible compared to $2D\tau$, the following is produced:

$$N_t = \frac{n}{f_n} \frac{\mu^2 E^2}{2D} = \frac{n}{f_n} \frac{\mu^2 V^2}{2DL^2}.$$ (G.3)

Note that the number of plates generated in a given time interval ($n\tau$) is linearly proportional to n as f_n is almost constant if $\alpha = 5$ and $n > 10$. (see Table G.1).

G.5 Height Equivalent of a Theoretical Plate

Plate height (H) is defined as the ratio of separation length and plate number (N). Using this and applying some elementary algebra to equation G.1 the following is obtained:

$$H = \frac{L_{tot}}{N} = \frac{f_n}{n} \frac{2D}{\mu E} = \frac{f_n}{n} \frac{2DL}{\mu V},$$ (G.4)

Table G.1 The minimum and maximum values of f_n, h_n, and g_n as a function of n (time $= n\tau$) and using $\alpha = 0.5$ (compression factor). The values in the limit of $n > 10$ are shown in the bottom line. The indexes "a" and "b" stand, respectively, for compression events just after and just before each detection. These are the limiting values that what will be observed, depending where the compression zones are positioned.

	Min	Max	Min	Max	Min	Max
n	f_n^b	f_n^a	h_n^a	h_n^b	g_n^b	g_n^a
1	0.250	1.000	0.577	1.000	1.000	2.000
2	0.313	1.250	0.775	1.342	0.947	1.894
3	0.328	1.313	0.882	1.528	0.922	1.845
4	0.332	1.328	0.939	1.627	0.909	1.818
5	0.333	1.332	0.969	1.679	0.900	1.801
6	0.333	1.333	0.984	1.705	0.895	1.789
7	0.333	1.333	0.992	1.719	0.891	1.781
8	0.333	1.333	0.996	1.725	0.887	1.775
9	0.333	1.333	0.998	1.729	0.885	1.770
10	0.333	1.333	0.999	1.730	0.883	1.766
>10	1/3	4/3	1	$\sqrt{3}$	$\sqrt{3}/2$	$\sqrt{3}$

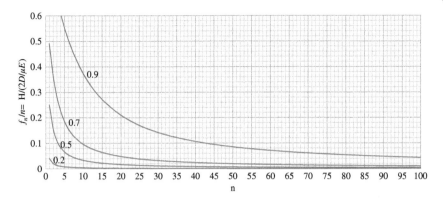

Figure G.2 Effect of the multiplying factor f_n/n on plate height (H). Four compression factors were used: $\alpha = 0.2, 0.5, 0.7$, and 0.9. Note that $\alpha = 0.2$ and 0.5 are very efficient in reducing the plate height (H).

The factor f_n/n is very efficient in reducing the plate height, as shown in Figure G.2. Here it is possible to see for $\alpha = 0.5$, f_n/n goes as $\simeq 1/n$ and f_n is almost constant. The smaller the compression factor α the narrower the plate height.

G.6 Resolution

Resolution can be calculated in the same way as shown in Section 4.9.2.1; however now the sum of the standard deviations is given by $(\sigma_i + \sigma_j) = \sqrt{2D_i \tau f_n} + \sqrt{2D_j \tau f_n}$. However, it is necessary to pay some attention in the calculation of Δx, as it is also affected by the compression events. This is calculated in Section G.10.2. The final expression of resolution in the presence of cyclic band compression events is given by:

$$R_{i,j} = h_n \frac{\Delta\mu}{4} \left(\frac{EL}{2\overline{D}\overline{\mu}} \right)^{\frac{1}{2}} = h_n \frac{\Delta\mu}{4} \left(\frac{V}{2\overline{D}\overline{\mu}} \right)^{\frac{1}{2}} \quad \text{where} \tag{G.5}$$

$$\left[\frac{\alpha(1 - \alpha^n)}{(1 - \alpha)(1 + \alpha^n)} \right]^{1/2} \leq h_n \leq \left[\frac{(1 + \alpha)(1 - \alpha^n)}{(1 - \alpha)(1 + \alpha^n)} \right]^{1/2}. \tag{G.6}$$

Note that h_n is almost independent of n when $\alpha = 0.5$ (see Table G.1). Therefore, cyclic band compression does not directly benefit this performance indicator. This happens because the band compression events, for instance with a compression factor of 0.5, reduce both the standard deviation of the bands and the inter-band distance by the same factor of 0.5, leaving the $\Delta x/\overline{\sigma}$ ratio unchangeable. Nevertheless, there are indirect benefits. Longer runs are possible because the standard deviations of the bands reach a steady state. This provides plenty of time to perform an array of new on-column assays, which were not possible previously (e.g., thermal stability studies, photostability or photorobustness measurements, among others).

G.7 Resolution per Unit Time

Dividing equation G.6 by the run time ($t = n\tau$) gives the resolution per unit time as a function of the operation parameters when a toroidal layout is used with cyclic band compression.

$$\frac{R_{i,j}}{t} = \frac{h_n}{n} \frac{\Delta\mu}{4} \left(\frac{\overline{\mu}E^3}{2\overline{D}L}\right)^{\frac{1}{2}} = \frac{h_n}{n} \frac{\Delta\mu}{4L^2} \left(\frac{\overline{\mu}V^3}{2\overline{D}}\right)^{\frac{1}{2}}. \tag{G.7}$$

This expression is almost independent of n and the same conclusions as found for equation G.6 also apply to this case. However, note that the right hand side expression is inversely proportional to L^2.

G.8 Band Capacity

Band capacity expresses the number of bands that can be separated and detected during a run. Only the bands that fit within two neighboring microholes ($L/2$) during each passage will be counted. In this case, at the mth detection the maximum number of bands that can be counted is given by the following expression:

$$B_m = \frac{L/2}{4\sigma} = \frac{1}{8} \left(\frac{\mu EL}{2D}\right)^{\frac{1}{2}} (f_m)^{-1/2}, \tag{G.8}$$

where, as before, $f_m = \alpha^2(1 - \alpha^{2m})/(1 - \alpha^2)$. Summing B_m from $m = 1$ to the last detection at $m = n$ gives the total number of bands detected in the time interval $n\tau$:

$$B = \frac{1}{8} \left(\frac{\mu EL}{2D}\right)^{\frac{1}{2}} \sum_{m=1}^{n} (f_m)^{-1/2},$$

$$= \frac{1}{8} \left(\frac{\mu V}{2D}\right)^{\frac{1}{2}} \sum_{m=1}^{n} (f_m)^{-1/2}. \tag{G.9}$$

This can be further simplified and produces what follows:

$$B = \frac{n\,g_n}{8} \left(\frac{\mu EL}{2D}\right)^{\frac{1}{2}} = \frac{n\,g_n}{8} \left(\frac{\mu V}{2D}\right)^{\frac{1}{2}}, \tag{G.10}$$

where g_n is a constant that depends on n and α only, as shown in Table G.1. Therefore, the number of analytes separated (or bands detected) grows linearly with time. This happens because the standard deviation of the bands reaches a steady state, as shown in Figure G.1, which in turn produces a constant number of bands at each detection.

G.9 Band Capacity per Unit Time

Dividing equation G.10 by the total run time ($t = n\tau$) gives the number of bands separated per unit time:

$$B_t = \frac{g_n}{8} \left(\frac{\mu^3 E^3}{2DL} \right)^{\frac{1}{2}} = \frac{g_n}{8L^2} \left(\frac{\mu^3 V^3}{2D} \right)^{\frac{1}{2}}. \tag{G.11}$$

This equation tells us that the number of bands separated per unit time is constant and does not depend on n. This is a consequence of the fact that, after a few turns, the standard deviation of the bands reaches a steady state and will indefinitely stay in a steady state.

The above performance indicators are confronted with the performance indicators of open and toroidal layout in Table G.2. The calculation of peak capacity and peak capacity per unit time is even more demanding and will be omitted in this book.

Notation used in Table G.2:

  (i) Tw/CBC: toroidal layout with cyclic band compression (a compression factor of $\alpha = 0.5$ and two compression events per turn are considered).
 (ii) The operational parameter n represents the number of detections or half-turns. The number of complete turns is given by $2n$.
(iii) The parameter $\sqrt{\overline{D}}$ is a symbol (not the square root of a mean value). Here is what it represents: $2\sqrt{\overline{D}} \equiv \sqrt{D_i} + \sqrt{D_j}$.
 (iv) $\overline{\mu}$ is the harmonic mean, $\overline{\mu} = 2\mu_i\mu_j/(\mu_i + \mu_j)$, and not the arithmetic mean since $t = (t_i + t_j)/2$ was used.

The functions f_n, h_n, and g_n are almost constant (see Table G.1) for $\alpha = 0.5$ (which was already demonstrated) [1,2]. As a consequence, the application of on-column cyclic band compression events (Tw/CBC) improves N, N_t, H, B, and B_t. Unfortunately, it does not improve R and R_t, as shown in Table G.2.

The number of plates (N) grows linearly with n in the toroidal layout and quadraticaly with n (as n^2) in the Tw/CBC. This happens because the standard deviation of the bands/peak reaches a steady state. The number of plates generated per unit time (N_t) is identical for both open and toroidal layouts, but for the Tw/CBC it is proportional to n because f_n is almost constant ($1/3 < f_n < 4/3$, as shown in Table G.1). Plate height is almost the same for the open and toroidal

Table G.2 Performance indicators of the open layout (open) compared with toroidal layout (toroidal) and toroidal with cyclic band compression (Tw/CBC). As stated in Table 5.4, they are expected to be valid for the three platforms (capillary, microchip, and slab) and at least six separation modes (AE, ELFSE, FSE, MEKC, MEEKC, and SE) where the cyclic on-column band compression is feasible.

	Open	Toroidal	Tw/CBC
$N =$	$\dfrac{\mu Vl}{2DL}$	$\dfrac{n\mu V}{2D}$	$\dfrac{n^2}{f_n}\dfrac{\mu V}{2D}$
$N_t =$	$\dfrac{\mu^2 E^2}{2D}$	$\dfrac{\mu^2 E^2}{2D}$	$\dfrac{n}{f_n}\dfrac{\mu^2 E^2}{2D}$
$H =$	$\dfrac{2D}{\mu E}$	$\dfrac{2D}{\mu E}$	$\dfrac{f_n}{n}\dfrac{2D}{\mu E}$
$R =$	$\dfrac{\Delta\mu}{4}\left(\dfrac{V}{2\bar\mu D}\right)^{\frac{1}{2}}$	$\dfrac{\Delta\mu}{4}\left(\dfrac{nV}{2\bar\mu D}\right)^{\frac{1}{2}}$	$h_n\dfrac{\Delta\mu}{4}\left(\dfrac{V}{2\bar\mu D}\right)^{\frac{1}{2}}$
$R_t =$	$\dfrac{\Delta\mu}{4}\left(\dfrac{\bar\mu E^3}{2\bar Dl}\right)^{\frac{1}{2}}$	$\dfrac{\Delta\mu}{4}\left(\dfrac{\bar\mu E^3}{2n\bar DL}\right)^{\frac{1}{2}}$	$\dfrac{h_n}{n}\dfrac{\Delta\mu}{4}\left(\dfrac{\bar\mu E^3}{2\bar DL}\right)^{\frac{1}{2}}$
$B =$	$\dfrac{1}{4}\left(\dfrac{\mu Vl}{2DL}\right)^{\frac{1}{2}}$	$\dfrac{2\sqrt{n}-1}{4}\left(\dfrac{\mu V}{2D}\right)^{\frac{1}{2}}$	$\dfrac{n\,g_n}{4}\left(\dfrac{\mu V}{2D}\right)^{\frac{1}{2}}$
$B_t =$	$\dfrac{1}{4}\left(\dfrac{\mu^3 E^3}{2Dl}\right)^{\frac{1}{2}}$	$\dfrac{1}{4}\left(\dfrac{2}{\sqrt{n}}-\dfrac{1}{n}\right)\left(\dfrac{\mu^3 E^3}{2DL}\right)^{\frac{1}{2}}$	$\dfrac{g_n}{8}\left(\dfrac{\mu^3 E^3}{2DL}\right)^{\frac{1}{2}}$

layouts; however in the Tw/CBC it decreases as $1/n$. The resolution (R) produced by the Tw/CBC is no better than that produced by the open layout; however it is proportional to $\sqrt{n}$ in the toroidal layout. The resolution produced per unit time (R_t) is almost the same for all layouts if $n = 1$; however in the toroidal layout (without cyclic band compression) it is smaller for larger values of n as it grows as $1/\sqrt{n}$ and in the Tw/CBC it decreases even faster, as $1/n$. This occurs because the inter band/peak distances are also affected by the successive compression events. Band capacity (B) is higher in the toroidal layouts than in the open layout as it is multiplied by a factor of $\sqrt{n}$ in the toroidal layout and by a factor of n in the Tw/CBC. Band capacity per unit time (B_t) is constant and independent of n in the Tw/CBC, no matter how many rounds the analytes complete. This happens because the cyclic band compression holds the bands and their variances in a steady state. However, the larger the value of n the smaller the band capacity per unit time will be in the toroidal layout. This occurs since the standard deviation of the bands increases with the square root of time, which allows successively fewer bands to fit in the length L.

Note that $\lim_{\alpha \to 1} f_n = n$ and $\lim_{\alpha \to 1} h_n = \sqrt{n}$. This converts the expressions valid for the Tw/CBC into the expressions of the toroidal layout. Moreover, the equations of the open layout are recovered by making $n = 1$ in the equations of the toroidal layout, which means one detection or one-half turn in the toroidal path.

G.10 Detailed Calculation of σ^2, Δx, f_n, and h_n

G.10.1 Peak Variance

G.10.1.1 Compression Events Before Each Detection

When no band compression events are applied then a given analyte migrates from $x = 0$ to $x = L$ with the velocity $v = \mu E$. The first compression zone extends from $x = L - \delta L$ to $x = L$ and the second compression zone extends from $x = 2L - \delta L$ to $x = 2L$. These are the zones located immediately before each detector. In this case, when the accessory that promotes band compression is active, the analyte migrates from $t = 0$ to $t = \tau - \delta\tau$ with a velocity of μE, and following this the analyte migrates with a velocity $\mu E \alpha$ for the remaining time ($\approx \delta\tau$) until it reaches $x = L$. At time $t = \tau - \delta\tau$ the standard deviation of the band will be $\sigma(\tau - \delta\tau) = \sqrt{2D(\tau - \delta\tau)}$ if $\sigma(0)$ is negligible compared to $2D\tau$. The band starts experiencing a compression event at $t = \tau - \delta\tau$ and its standard deviation and variance at time $t = \tau$ are given by:

$$\sigma(\tau) = \sqrt{2D\tau}\alpha \quad \text{and} \quad \sigma^2(\tau) = 2D\tau\alpha^2. \tag{G.12}$$

If the initial injected sample plug variance, $\sigma^2(0)$, has to be taken into account, then $\sigma^2(\tau) = [\sigma^2(0) + 2D\tau]\alpha^2$. At time $t = 2\tau - \delta\tau$ the variance will be $\sigma^2(2\tau - \delta\tau) = [\sigma^2(0) + 2D\tau]\alpha^2 + 2D(\tau - \delta\tau)$, and at time $t = 2\tau$ it will be $\sigma^2(2\tau) = (\sigma^2(0) + 2D\tau)\alpha^4 + 2D\tau\alpha^2$. Therefore, the variance of the nth detection, which will occur at time $t \approx n\tau$, is given by:

$$\sigma^2(n\tau) = \sigma^2(0)\alpha^{2n} + 2D\tau \sum_{m=1}^{n} \alpha^{2m},$$

$$= \sigma^2(0)\alpha^{2n} + 2D\tau f_n^b, \quad \text{where}$$

$$f_n^b = \frac{\alpha^2(1 - \alpha^{2n})}{1 - \alpha^2}. \tag{G.13}$$

Note that f_n^b only depends on α and n. The label "b" indicates that the compression events were performed immediately before each detection, as shown in Section G.10.1.2. Table G.1 gives some values of f_n^b for when $\alpha = 0.5$.

If there are other time dependent band broadening mechanisms whose variances are additive to $2Dt$ [10], then their dispersion coefficients can be added to the constant D in equation G.13. Examples of such dispersion coefficients (representing band broadening mechanisms) are: D_{ads} (wall adsorption), D_{coil} (coiling of the separation track), D_{emd} (electromigration dispersion), D_{ieof} (inhomogeneous EOF), D_{pvf} (Poiseuille velocity flow), and D_{temp} (temperature gradients) (see Section 2.13).

G.10.1.2 Compression Events After Each Detection

When compression events are performed just after each detection, a similar expression can be deduced using the procedure detailed in the previous subsection.

At time $t = \tau$ the variance will be $\sigma^2(\tau) = \sigma^2(0)\alpha^2 + 2D(\tau)$ because the first compression event occurred between $x = 0$ and $x = \delta L$. At $t = \tau + \delta\tau$, just after crossing the second compression zone, the variance will be given by $\sigma^2(\tau + \delta\tau) = [\sigma^2(0)\alpha^2 + 2D(\tau)]\alpha^2$ because the compression events happen immediately after each detection. At time $t = 2\tau$ the variance will be given by $\sigma^2(2\tau) = \sigma^2(0)\alpha^4 + 2D\tau\alpha^2 + 2D\tau$. At time $n\tau$ the variance will be given by:

$$\sigma^2(n\tau) = \sigma^2(0)\alpha^{2n} + 2D\tau \sum_{m=1}^{n} \alpha^{2m-2},$$

$$= \sigma^2(0)\alpha^{2n} + 2D\tau f_n^a, \quad \text{where}$$

$$f_n^a = \frac{1 - \alpha^{2n}}{1 - \alpha^2}. \tag{G.14}$$

The label "a" is used to remind users that the compression events are performed just after each detection. Note that $f_n^b < f_n^a$ because $0 < \alpha \le 1$. An intermediate value of f_n lying between f_n^b and f_n^a would be expected if the compression events are performed at intermediate (but opposite) points along the toroid. Table G.1 shows the values of f_n^b and f_n^a as a function of n when $\alpha = 0.5$.

Figure G.1 shows how the variance changes over time when one cyclic band compression event is performed before each detection (each detection is represented by n in the graph). Note that two detections are performed per turn and therefore two compression events occur per turn. The graph shows three distinct initial variances: $\sigma^2(0) = 0$, $\sigma^2(0)/2D\tau = 4$, and $\sigma^2(0)/2D\tau = 9$. Two compression factors were used: $\alpha = 0.8$ and $\alpha = 0.5$ (the latter has already been demonstrated).[20,21] Note that the variance reaches a steady state after four completed turns (following the $n = $ 8th detection) when $\alpha = 0.5$ and after eleven completed turns (following the $n = $ 22nd detection) when $\alpha = 0.8$.

G.10.2   Inter-Peak Spacing (Δx)

G.10.2.1   Compression Events Before Each Detection

Before the first compression at $t = \tau - \delta\tau$, when the two bands have just reached the compression zone, the distance between the centers of mass of the bands will be: $\Delta x(\delta\tau) = x_i - x_j = \Delta\mu E(\tau - \delta\tau)$. However, the distance between the centers of mass of the bands at the time of the first detection (τ) will be $\Delta x(\tau) = \Delta\mu E\tau\alpha$, since the band compression event is performed just before each detection. Figure 2.21 illustrates this effect, i.e. the reduction of the inter-band distance. At the end of the second compression Δx will be: $\Delta x(2\tau) = (\Delta\mu E\tau\alpha^2 + \Delta\mu E\tau)\alpha$. This means that after the nth detection the value of Δx will be:

$$\Delta x(n\tau) = \Delta\mu E\tau \sum_{m=1}^{n} \alpha^m,$$

$$= \Delta\mu E\tau j_n^b, \quad \text{where}$$

$$j_n^b = \frac{\alpha(1 - \alpha^n)}{1 - \alpha}. \tag{G.15}$$

G.10.2.2   Compression Events After Each Detection

When the compression events occur just after each detection the distance between the centers of mass of the bands at time τ will be: $\Delta x(\tau) = \Delta\mu E\tau$. This happens because the compression event performed at time $t = \delta\tau$ has a negligible effect as $\Delta x(0) \simeq 0$. At time $\tau + \delta\tau$ the value of Δx will be $\Delta x(\tau + \delta\tau) = \Delta\mu E(\tau + \delta\tau)\alpha$. At time 2τ the value of Δx will be $\Delta x(2\tau) = \Delta\mu E\tau\alpha + \Delta\mu E\tau$. At time $n\tau$ it will be:

$$\Delta x(n\tau) = \Delta\mu E\tau \sum_{m=1}^{n} \alpha^{m-1},$$

$$= \Delta\mu E\tau j_n^a, \quad \text{where}$$

$$j_n^a = \frac{(1 - \alpha^n)}{1 - \alpha}. \tag{G.16}$$

The expression of equation G.16 gives the maximum value of Δx that can be expected at each detection, since the compression events are performed just after each detection.

G.10.3   Calculation of the Values of h_n

The expression of $h_n = j_n / \sqrt{f_n}$ appears in the calculation of resolution (R) when cyclic band compression events are applied (see Section G.6). Here the function h_n, which is a function of n and α, is adopted to simplify the notation. Again, two limiting values for h_n will be found: a minimum and a maximum. However, these limiting values must first be calculated and then compared to see which is the minimum and which is the maximum.

G.10.3.1 Compression Events Before Each Detection

When the compression events are performed immediately before each detection, h_n^b must be calculated using the following expressions:

$$j_n^b = \frac{\alpha(1 - \alpha^n)}{1 - \alpha} \quad \text{and} \tag{G.17}$$

$$f_n^b = \frac{\alpha^2(1 - \alpha^{2n})}{1 - \alpha^2}. \tag{G.18}$$

After some calculations $h_n^b = j_n^b / \sqrt{f_n^b}$ becomes:

$$h_n^b = \left[\frac{\alpha(1 - \alpha^n)}{(1 - \alpha)(1 + \alpha^n)} \right]^{1/2}. \tag{G.19}$$

G.10.3.2 Compression Events After Each Detection

When the compression events are performed just after each detection, the following expressions must be used:

$$j_n^a = \frac{(1 - \alpha^n)}{1 - \alpha} \quad \text{and} \tag{G.20}$$

$$f_n^a = \frac{(1 - \alpha^{2n})}{1 - \alpha^2}. \tag{G.21}$$

After some calculations $h_n^a = j_n^a / \sqrt{f_n^a}$ becomes:

$$h_n^a = \left[\frac{(1 + \alpha)(1 - \alpha^n)}{(1 - \alpha)(1 + \alpha^n)} \right]^{1/2}. \tag{G.22}$$

Comparison of equations G.19 and G.22 shows that h_n ranges between the following limits:

$$\left[\frac{\alpha(1 - \alpha^n)}{(1 - \alpha)(1 + \alpha^n)} \right]^{1/2} \leq h_n \leq \left[\frac{(1 + \alpha)(1 - \alpha^n)}{(1 - \alpha)(1 + \alpha^n)} \right]^{1/2}. \tag{G.23}$$

Note that two detectors and two compression zones are considered in this book. This is the ideal setup if cyclic band compression events are applied in an EST with a toroidal layout, four microholes, and four reservoirs. The problem of an arbitrary number of compression zones will not be described here, but it can be found elsewhere [11].

References

1 Mandaji, M., Rübensam, G., Hoff, R.B. et al. (2009). *Electrophoresis* 30: 1501–1509.

2 Mandaji, M., Rübensam, G., Hoff, R.B. et al. (2009). *Electrophoresis* 30: 1510–1515.

3 Malá, Z., Křivánková, L., Gebauer, P., and Boček, P. (2007). *Electrophoresis* 28: 243–253.

4 Malá, Z., Šlampová, A., Gebauer, P., and Boček, P. (2009). *Electrophoresis* 30: 215–229.

5 Malá, Z., Gebauer, P., and Boček, P. (2011). *Electrophoresis* 32: 116–126.

6 Šlampová, A., Malá, Z., Pantčková, P. et al. (2013). *Electrophoresis* 34: 3–18.

7 Malá, Z., Šlampová, A., Křivánková, L. et al. (2015). *Electrophoresis* 36: 15–35.

8 Šlampová, A., Malá, Z., Gebauer, P., and Boček, P. (2017). *Electrophoresis* 38: 20–32.

9 Šlampová, A., Malá, Z., and Gebauer, P. (2019). *Electrophoresis* 40: 40–54.

10 Virtanen, R. (1993). *Electrophoresis* 14: 1266–1270.

11 Kist, T.B.L. (2018). *Journal of Separation Science* 41: 2640–2650.

Index

*Open and Toroidal Electrophoresis: Ultra-High Separation Efficiencies in Capillaries,
Microchips, and Slabs,* First Edition. Tarso B. Ledur Kist.
© 2020 John Wiley & Sons Ltd. Published 2020 by John Wiley & Sons Ltd.